Aluminum Alloys and Aluminum-Based Matrix Composites

Aluminum Alloys and Aluminum-Based Matrix Composites

Editors

Di Feng
Qianhao Zang
Ying Liu
Yunsoo Lee

Basel • Beijing • Wuhan • Barcelona • Belgrade • Novi Sad • Cluj • Manchester

Editors

Di Feng
Jiangsu University of Science
and Technology
Zhenjiang, China

Qianhao Zang
Jiangsu University of Science
and Technology
Zhenjiang, China

Ying Liu
Nanjing University of Science
and Technology
Nanjing, China

Yunsoo Lee
Korea Institute of Materials
Science (KIMS)
Changwon, Korea

Editorial Office
MDPI
St. Alban-Anlage 66
4052 Basel, Switzerland

This is a reprint of articles from the Special Issue published online in the open access journal *Metals* (ISSN 2075-4701) (available at: https://www.mdpi.com/journal/metals/special_issues/9NC19AE2CA).

For citation purposes, cite each article independently as indicated on the article page online and as indicated below:

Lastname, A.A.; Lastname, B.B. Article Title. *Journal Name* **Year**, *Volume Number*, Page Range.

ISBN 978-3-0365-9612-9 (Hbk)
ISBN 978-3-0365-9613-6 (PDF)
doi.org/10.3390/books978-3-0365-9613-6

© 2023 by the authors. Articles in this book are Open Access and distributed under the Creative Commons Attribution (CC BY) license. The book as a whole is distributed by MDPI under the terms and conditions of the Creative Commons Attribution-NonCommercial-NoDerivs (CC BY-NC-ND) license.

Contents

About the Editors vii

Di Feng, Rui Xu, Jichen Li, Wenjie Huang, Jingtao Wang, Ying Liu, et al.
Microstructure Evolution Behavior of Spray-Deposited 7055 Aluminum Alloy during Hot Deformation
Reprinted from: *Metals* 2022, 12, 1982, doi:10.3390/met12111982 1

Yi Tao, Yonghui Wang, Qiang He, Daoming Xu and Lizheng Li
Comparative Study and Multi-Objective Crashworthiness Optimization Design of Foam and Honeycomb-Filled Novel Aluminum Thin-Walled Tubes
Reprinted from: *Metals* 2022, 12, 2163, doi:10.3390/met12122163 19

Pengfei Zhou, Dongtao Wang, Hiromi Nagaumi, Rui Wang, Xiaozu Zhang, Xinzhong Li, et al.
Microstructural Evolution and Mechanical Properties of Al-Si-Mg-Cu Cast Alloys with Different Cu Contents
Reprinted from: *Metals* 2023, 13, 98, doi:10.3390/met13010098 35

Guang Chen, Changcai Zhao, Haiwei Shi, Qingxing Zhu, Guoyi Shen, Zheng Liu, et al.
Research on the 2A11 Aluminum Alloy Sheet Cyclic Tension–Compression Test and Its Application in a Mixed Hardening Model
Reprinted from: *Metals* 2023, 13, 229, doi:10.3390/met13020229 49

Junlong Tang, Shenbo Liu, Dongxue Zhao, Lijun Tang, Wanghui Zou and Bin Zheng
An Algorithm for Real-Time Aluminum Profile Surface Defects Detection Based on Lightweight Network Structure
Reprinted from: *Metals* 2023, 13, 507, doi:10.3390/met13030507 69

Jiansheng Xia, Rongtao Liu, Jun Zhao, Yingping Guan and Shasha Dou
Study on Friction Characteristics of AA7075 Aluminum Alloy under Pulse Current-Assisted Hot Stamping
Reprinted from: *Metals* 2023, 13, 972, doi:10.3390/met13050972 83

Haihao Teng, Yufeng Xia, Chenghai Pan and Yan Li
Modified Voce-Type Constitutive Model on Solid Solution State 7050 Aluminum Alloy during Warm Compression Process
Reprinted from: *Metals* 2023, 13, 989, doi:10.3390/met13050989 103

Jialong Kang, Yaoran Cui, Dapeng Zhong, Guibao Qiu and Xuewei Lv
A New Method for Preparing Titanium Aluminium Alloy Powder
Reprinted from: *Metals* 2023, 13, 1436, doi:10.3390/met13081436 119

Lihua Zhang, Jijun Li, Jing Zhang, Yanjie Liu and Lin Lin
First-Principle Investigation into Mechanical Properties of $Al_6Mg_1Zr_1$ under Uniaxial Tension Strain on the Basis of Density Functional Theory
Reprinted from: *Metals* 2023, 13, 1569, doi:10.3390/met13091569 135

Puli Cao, Guilan Xie, Chengbo Li, Daibo Zhu, Di Feng, Bo Xiao and Cai Zhao
Investigation of the Quenching Sensitivity of the Mechanical and Corrosion Properties of 7475 Aluminum Alloy
Reprinted from: *Metals* 2023, 13, 1656, doi:10.3390/met13101656 147

Di Feng, Qianhao Zang, Ying Liu and Yunsoo Lee
Aluminum Alloys and Aluminum-Based Matrix Composites
Reprinted from: *Metals* 2023, 13, 1870, doi:10.3390/met13111870 161

About the Editors

Di Feng

Di Feng is an associate professor at the School of Materials Science and Engineering at Jiangsu University of Science and Technology, China, with nearly 20 years of experience in theoretical research and engineering application research. He has published over 50 scientific papers in international/national journals and has also published one monograph. His main research areas are the design of aluminum alloys with new chemical compositions and the development of new forming techniques. He has recently developed an interest in the recycling technology of aluminum materials, achieving material conservation, and recycling through the recycling of cutting residues.

Qianhao Zang

Qianhao Zang is an associate professor at the School of Materials Science and Engineering at Jiangsu University of Science and Technology in Zhenjiang, China. He has more than 10 years of experience in research on aluminum and magnesium alloys. He has published more than 30 scientific papers in international journals. His main research areas are the hot deformation behavior, microstructure, and mechanical properties of aluminum alloys. He has recently become interested in the microstructure and mechanical properties of aluminum alloys fabricated by Twin roll casting.

Ying Liu

Ying Liu is an associate professor at the School of Materials Science and Engineering at Nanjing University of Science and Technology, China. She has published more than 50 scientific papers in international/national journals. Her main research area includes the control of the deformation structure of metals and alloys (including recrystallization structures and the regulation of the precipitated second phase), research on the nanotechnology of material surfaces, and the failure analysis of materials.

Yunsoo Lee

Yunsoo Lee is a senior researcher at the Department of Aluminum at the Korea Institute of Materials Science (KIMS) in Changwon, South Korea. He has more than 10 years of experience in research on aluminum materials. He has published more than 40 scientific papers in international/national journals and conference proceedings. His main research areas are alloy design through thermodynamic calculations of aluminum alloys, continuous casting, plastic deformation, heat treatment, and porous metallic materials. He has recently become interested in the circular economy through the recycling of by-products and scraps generated from the metallic material manufacturing.

Article

Microstructure Evolution Behavior of Spray-Deposited 7055 Aluminum Alloy during Hot Deformation

Di Feng [1,2,3,*], Rui Xu [1], Jichen Li [1], Wenjie Huang [1], Jingtao Wang [2], Ying Liu [2,*], Linxiang Zhao [3], Chengbo Li [4] and Hao Zhang [5]

1. Department of Materials Science and Engineering, Jiangsu University of Science and Technology, Zhenjiang 212003, China
2. Department of Materials Science and Engineering, Nanjing University of Science and Technology, Nanjing 210094, China
3. Suzhou Feihua Aluminum Industry Co., Ltd., Suzhou 215164, China
4. Department of Machinery and Mechanics, Xiangtan University, Xiangtan 411105, China
5. Jiangsu Haoran Spray Forming Alloy Co., Ltd., Zhenjiang 212009, China
* Correspondence: difeng1984@just.edu.cn (D.F.); liuying517@njust.edu.cn (Y.L.); Tel.: +86-0511-8840-1184 (D.F.)

Abstract: The evolution behaviors of the second phase, substructure and grain of the spray-deposited 7055 aluminum alloy during hot compression at 300~470 °C were studied by scanning electron microscopy (SEM), electron backscatter diffraction (EBSD) and transmission electron microscopy (TEM). Results show that the AlZnMgCu phase resulting from the deposition process dissolves gradually with the increase in deformation temperature, but the Al_7Cu_2Fe phase remains unchanged. The plastic instability of the spray-deposited 7055 aluminum alloy occurs at 470 °C with a 1~5 s^{-1} strain rate range. Partial dynamic recrystallization (PDRX) adjacent to the original high angle grain boundaries (HAGBs) not only occurs at 300~400 °C with the low strain rates ranging from 0.001 to 0.1 s^{-1} but also at 450 °C with a high strain rate of 5 s^{-1}. Continuous dynamic recrystallization (CDRX) appears at 450 °C with a low strain rate of 0.001 s^{-1}. The primary nucleation mechanism of PDRX includes the rotation of the subgrain adjacent to the original HAGBs and the subgrain boundary migration. The homogeneous misorientation increase in subgrains is the crucial nucleation mechanism of CDRX. At 300~400 °C, the residual coarse particle stimulated (PSN) nucleation can also be observed.

Keywords: spray deposition; 7055 aluminum alloy; dynamic recovery; dynamic recrystallization

1. Introduction

Aluminum alloys are widely used in preparing aerospace structural parts due to their high specific strength, excellent corrosion resistance and machinability [1–3]. The Zn/Mg ratio, Zn/Cu ratio and element uniform distribution are the crucial factors determining the service performance of the Al-Zn-Mg-Cu alloy. However, the semi-continuous cast 7055 aluminum alloy has severe composition segregation and a casting cracking tendency [4,5]. Therefore, the primary alloying element content of large-size ingots is limited in the 7055 aluminum alloy. For example, the Zn element is generally set to the lower level of the nominal composition, which restricts the application potential of the Al-Zn-Mg-Cu alloy [6].

Spray deposition is an advanced ingot-forming technology based on rapid solidification [7,8]. In this process, the metal solution is atomized into molten droplets by high-pressure inert gas first. Then, the droplets sputter to a deposition disk at the bottom to form one solidification layer. Finally, a cylindrical ingot can be obtained by stacking layer by layer. During the flight of droplets toward the deposition disc, the cooling rate can reach 10^3 °C/s. Therefore, the dendrite structure is eliminated, and the grain is extraordinarily refined (20–30 μm) by the extreme undercooling degree. More importantly, the alloying elements in the spray-deposited 7055 aluminum alloy have a significant promotion, and there is no macroscopic segregation [9–11]. Therefore, the spray-deposited 7055 aluminum alloy has an extensive application potential for aerospace products.

It should be pointed out that the large size spray-deposited ingot needs to be densified by hot extrusion due to the presence of a certain proportion of "pores" [12,13]. Densification realized by extrusion is also the forming technology of plates and profiles [13,14]. It is well known that the controlling of the recrystallization degree and recrystallized grain size is one of the critical factors in ensuring the mechanical properties and corrosion resistance of the Al-Zn-Mg-Cu alloy. The hot extrusion temperature, extrusion rate and extrusion ratio directly determine the characteristic of the deformation substructure or recrystallized microstructure. In order to bring out the latent potentialities of the finer microstructure and higher element content, it is necessary to investigate the thermal deformation behavior of the spray-deposited 7055 aluminum alloy.

Feng [15] investigated the hot deformation behavior of a Al–7.68 Zn–2.12 Mg–1.98 Cu–0.12 Zr alloy produced by semi continuous casting with a homogenization state. The results revealed that the primary dynamic recrystallization (DRX) mechanism is "the bowing of original grain boundary". The effect of grain size inhomogeneity of the ingot on the dynamic softening behavior of the 7055 aluminum alloy was also researched in detail [16,17]. As mentioned above, the spray-deposited 7055 aluminum alloy showed a finer microstructure, indicating distinctive dynamic softening behaviors. Luo [2] studied the microstructure evolution of the spray-deposited and as-extruded 7055 aluminum alloy during hot compression. However, a detailed and systematic study of the hot deformation behavior, especially the DRX behavior of the as-sprayed 7055 aluminum alloy, has not been reported before.

In this paper, the evolution behavior of the second phase and the grain of the spray-deposited 7055 aluminum alloy was studied by the hot compression test. The dynamic softening mechanism of the rapidly solidified alloy was discussed. Results are also of great significance for the understanding of the thermal deformation behavior and the hot working process optimization of other spray-deposited aluminum alloys.

2. Materials and Methods

The actual composition (mass fraction, %) of the spray-deposited 7055 aluminum alloy ingot is: Zn 8.32, Mg 2.10, Cu 2.24, Zr 0.12, Si 0.030, Fe 0.045 and Al allowance. The specimens of 15 mm in height and 10 mm in diameter were machined from a spray-deposited ingot. The 0.2 mm deep grooves on the end faces were also machined. During compression, the grooves were filled with lubricant to reduce friction. A hot compression test was carried out on the Gleeble-3500 thermal simulator. The specimens were heated to 300 °C, 350 °C, 400 °C, 450 °C and 470 °C within 3 min, respectively, and held for 3 min prior to compression. Then, the specimens were compressed to the required reduction of 60% with the strain rates of $0.001\ s^{-1}$, $0.01\ s^{-1}$, $0.1\ s^{-1}$, $1\ s^{-1}$ and $5\ s^{-1}$, respectively. Water cooling was conducted immediately after hot compression to freeze the deformed microstructure [18,19].

The grains and dislocations were observed by the TecnaiG^2 20 transmission electron microscope (TEM) and field emission gun scanning electron microscope (SEM) equipped with an electron backscatter diffraction (EBSD) system. The TEM and EBSD specimens were thin foils with the thickness of 80 μm and the diameter of 3 mm, which were electropolished in a solution containing 30%HNO$_3$ in methanol at ~−25 °C and 15~20 V [17]. On account of the detection limitation in the hot worked structure, boundaries with a misorientation angle of less than 2° were not taken into account for EBSD observation. The EBSD test step size varies from 0.75 to 1.3 μm depending on the substructure scale [20]. The X-ray diffraction (XRD) experiments for the spray deposited specimen and the compressed one were performed on a XRD-6000 X-ray diffractometer using Cu Kα radiation. The measurement step is 1°/min, and the scanning range is 15~90° [21]. MDI Jade 6.5 was used for the qualitative analysis.

3. Results

3.1. True Stress—True Strain Curve

The true stress-strain curves of the spray-deposited 7055 aluminum alloy are shown in Figure 1. Results show that the flow stress increases with the decrease in the deformation temperature or the increase in the strain rate. The steady-state flow stress is only about 10 MPa at 470 °C, 0.001 s^{-1}, while the peak stress can reach 140 MPa at 300 °C, 5 s^{-1}. In addition, the flow stress increases rapidly at the beginning of compression (true strain < 0.1). When the true strain exceeds 0.1, the growth rate of flow stress decreases gradually.

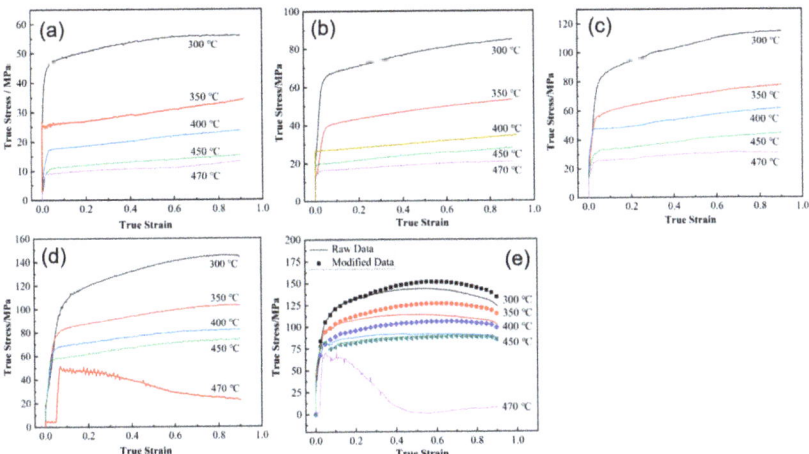

Figure 1. True stress-true strain curves under different strain rate and temperature conditions (Data are corrected based on deformation temperature rise effects for 5 s^{-1} compressed samples) (a) 0.001 s^{-1}; (b) 0.01 s^{-1}; (c) 0.1 s^{-1}; (d) 1 s^{-1}; (e) 5 s^{-1}.

Under the condition of a low deformation temperature (≤350 °C), it reveals that the flow stress keeps an upward trend with the increase in strain. Taking the hot compression temperature of 300 °C as an example, it can be seen that, from the strain of 0.05 to the completion of hot compression deformation, the flow stress increments ($\Delta\sigma$) at different strain rates are about 10 MPa (0.001 s^{-1}), 20 MPa (0.01 s^{-1}), 35 MPa (0.1 s^{-1}) and 40 MPa (1 s^{-1}), respectively. However, at intermediate and low strain rates (≤1s^{-1}) and temperatures above 400 °C, the stress levels are almost unchanged ($\Delta\sigma$ <10 MPa) after the rapid work hardening. Especially when the deformation temperature reaches 470 °C, the flow stress increment is negligible when the strain is larger than 0.05. It can be concluded that the spray-deposited 7055 aluminum alloy is also sensitive to the deformation temperature and strain rate, and its rheological behavior is more susceptible to the deformation temperature.

The stress-strain curves corresponding to 470 °C and intermediate (1 s^{-1}) and high (≥5 s^{-1}) strain rates decrease sharply after reaching the peak stress. In the process of high-speed thermal deformation, the external input mechanical energy accumulates rapidly by dislocation multiplication. The DRX behavior can effectively reduce the dislocation density and avoid the material fracture caused by severe work hardening. However, at a higher deformation temperature (470 °C), the temperature rise caused by the high strain rate exceeds the melting point of eutectic structures on the grain boundary, which eventually leading to the intergranular fracture. Therefore, it is preliminarily concluded that the deformation temperature of the spray-deposited 7055 aluminum alloy should not exceed 450 °C in the rapid deformation process, such as forging.

3.2. Deformation Microstructure Analysis

3.2.1. The Second Phase

Figure 2 shows the microstructure of as-sprayed and as-compressed states, respectively. It can be seen that the volume fraction of the white coarse second phase in the as-deposited specimen is the highest. The length-thickness ratio of the second phase inside the grain is small and evenly distributed. However, the ones on the grain boundary have a larger length-thick ratio. The coarse phases distribute along the grain boundary, outlining the morphology of the equiaxed grain (Figure 2a). Besides, the arrangement of the second phase in the as-sprayed specimen has no directivity.

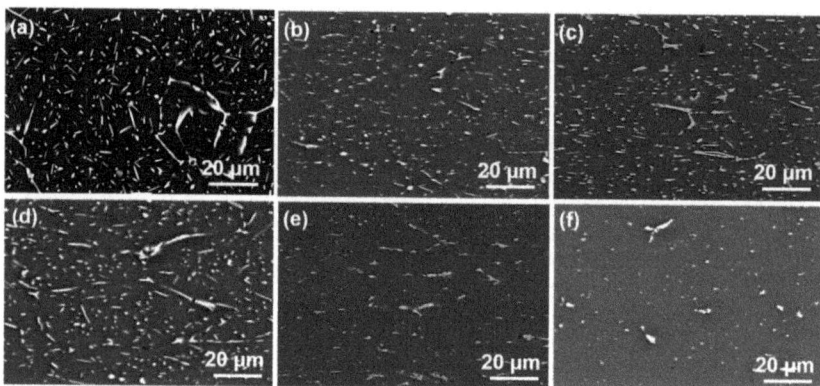

Figure 2. SEM images of 7055 aluminum alloy under spray-deposited state (a), 300 °C/5 s^{-1} (b), 300 °C/0.001 s^{-1} (c), 400 °C/0.1 s^{-1} (d), 450 °C/5 s^{-1} (e) and 450 °C/0.001 s^{-1} (f).

With the increase in the hot compression temperature or the decrease in the strain rate, the volume fraction of the coarse second phase decreases gradually. Image J software was used to calculate the area fraction of the second phase. Results show that, after hot compression at 300 °C (Figure 2b,c), the area fraction of the second phase is about 13%. The number of the intragranular phase decreases significantly, while the grain boundary phase remains unchanged. When the deformation temperature is 400 °C, the total area fraction decreases to about 8% (Figure 2d). The length-thickness ratio also reduces significantly. When the deformation temperature rises to 450 °C, the area fraction of the second phase in the specimen deformed at 5 s^{-1} is only about 2% (Figure 2e). However, the second phase in the specimen deformed at 0.001 s^{-1} almost dissolves completely. The residual phases on the grain boundary also spheroidize and coarsen (Figure 2f).

The element plane scanning results under different deformation conditions are shown in Figure 3. It reveals that the residual second phases are mainly AlZnMgCu quaternary particles (Figure 3a–e). Fe-containing phases also exist (Figure 3a,c). The type of the second phase remaining after hot compression is consistent with that of the as-deposited state. The Fe-containing phase is the crystalline phase formed in the deposition process and cannot be dissolved by heat treatment or thermal deformation.

The literature [22] shows that when the deformation temperature rises to 400~450 °C, the Fe-containing phase is exposed due to the dissolution of the associated AlZnMgCu phase. This coarse second phase will hinder the dislocation movement during thermal deformation, thus affecting the evolution behaviors of the substructure or the migration of high angle grain boundaries (HAGBs) during recrystallization.

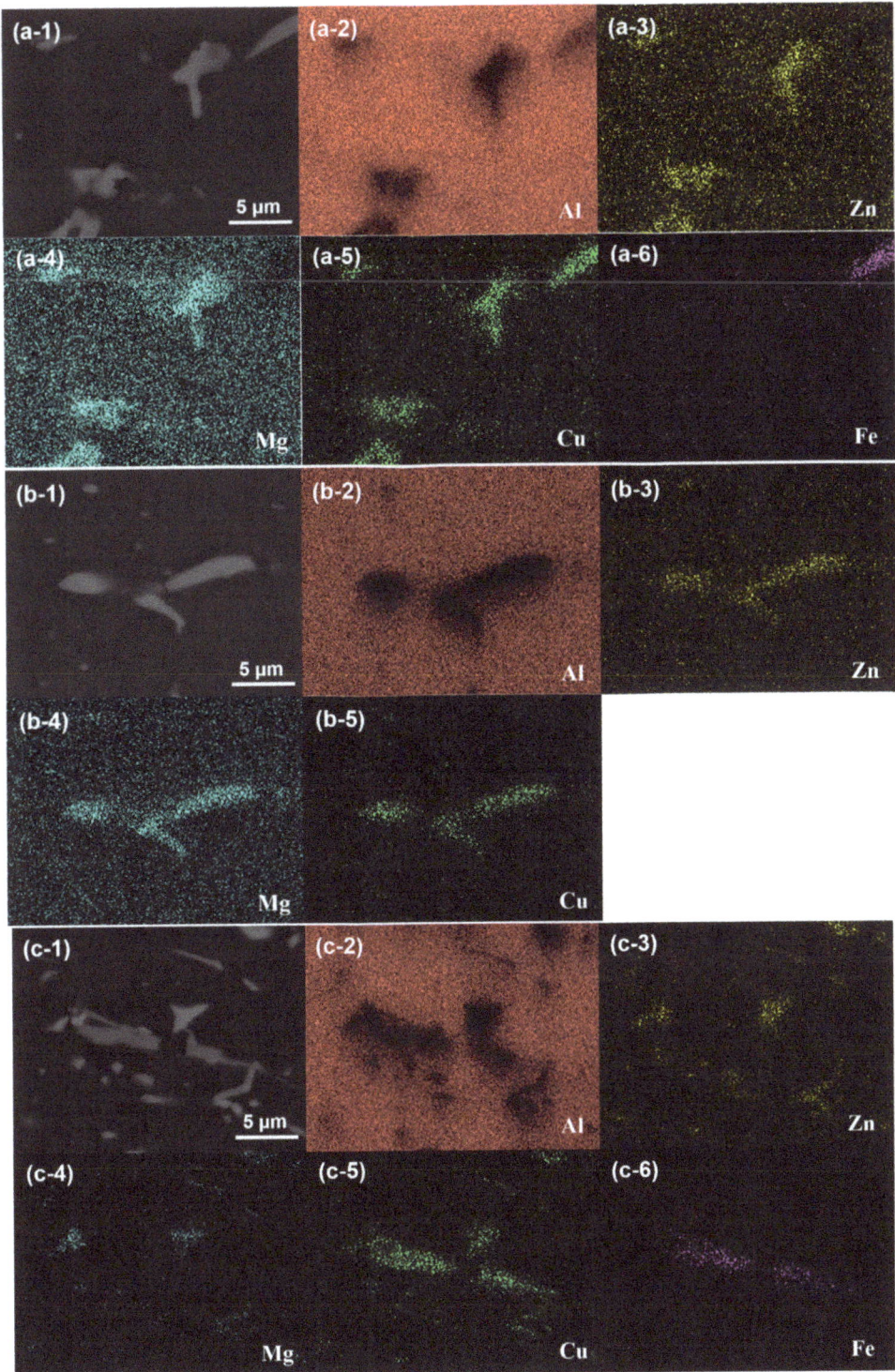

Figure 3. Cont.

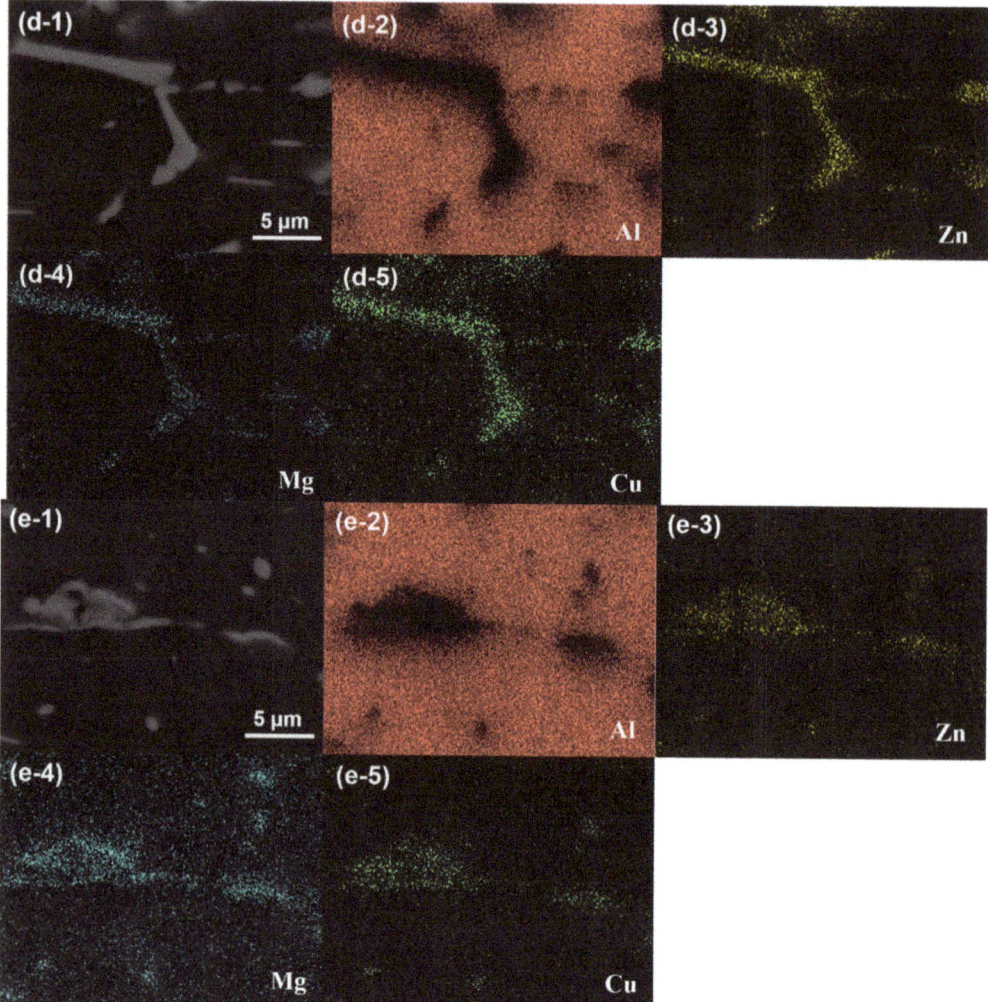

Figure 3. Element plane scanning images of the second phase of spray-deposited 7055 aluminum alloy under hot compression conditions of 450 °C/0.001 s^{-1} (**a**), 450 °C/5 s^{-1} (**b**), 400 °C/0.1 s^{-1} (**c**), 300 °C/0.001 s^{-1} (**d**) and 300 °C/5 s^{-1} (**e**).

The XRD patterns of the spray-deposited specimen and the one hot compressed under 450 °C/0.001s^{-1} are shown in Figure 4. It is depicted that the AlZnMgCu phase has the same crystal structure as MgZn$_2$ [23]. Based on EDS, we know the second phase also contains Al and Cu. So, the AlZnMgCu phase can also be defined as the Mg (Zn,Cu,Al)$_2$ phase [24]. Besides, the AlZnMgCu phase or Mg (Zn,Cu,Al)$_2$ phase decreases with the increase in the compressed temperature and the decrease in the strain rate, which is consistent with the SEM analysis results.

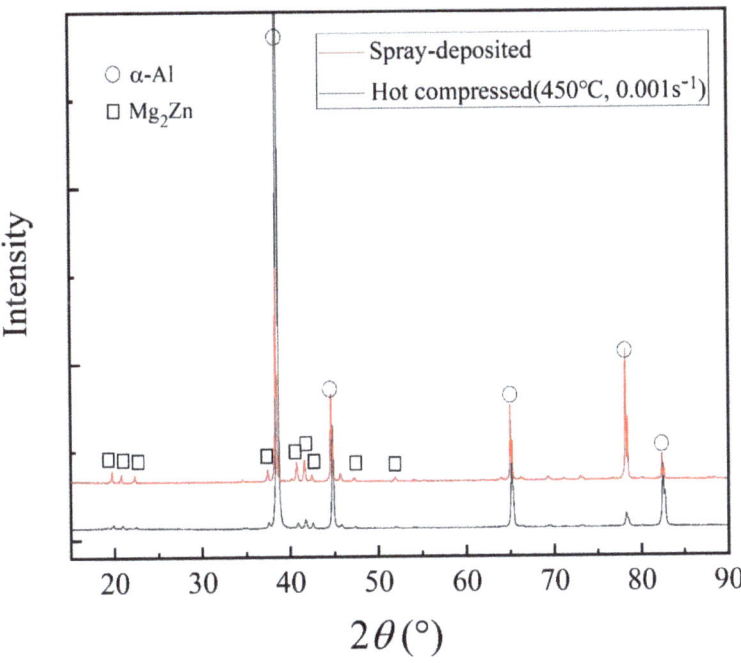

Figure 4. The XRD patterns of the spray-deposited specimen and the one deformed under 450 °C/0.001 s^{-1}, respectively.

3.2.2. Grain Morphology

Figure 5 shows the EBSD images of the hot compressed alloy before and after compression. The grains of the as-spray-deposited state are equiaxed (as shown in Figure 5a), and the size ranges from 20 to 50 μm. After deformed at the low temperature of 300 °C and a high speed of 5 s^{-1} (as shown in Figure 5b), the original equiaxed grain is compressed into a flat shape with a high-density deformation substructure inside. A small number of fine equiaxed recrystallized grains with a size range of only 3~5 μm can be observed at the original HAGBs. When the strain rate decreases to 0.001s^{-1} (Figure 5c), the size of recrystallized grains at HAGBs increases to 5~10 μm. The average size of the deformed substructure increases to about 10 μm. When the deformation temperature increases to 400 °C, the size of DRX grains and deformation substructures continue increasing (> 10 μm), as shown in Figure 5d. At the deformation temperature of 450 °C, the fine DRX grains distributed continuously on the grain boundaries disappear. Equiaxed DRX grains with the large size range of 10~50 μm can be observed at the trigeminal grain boundaries (Figure 5e). The volume fraction of the substructure decreased significantly, and some subgrains even grew to a scale of 50 μm (Figure 5f).

According to the above analysis, the primary dynamic softening mechanism of the spray-deposited 7055 aluminum alloy under low temperatures is dynamic recovery (DRV). The limited DRX microstructure is characterized by small-size equiaxed grains with a pearl necklace shape at the original HAGBs. Therefore, the dynamic softening effect under a low temperature is limited. When the dislocation multiplication is quicker than annihilation, the flow stress will continue rising slowly, as shown in the stress-strain curve under the 300 °C deformation condition in Figure 1.

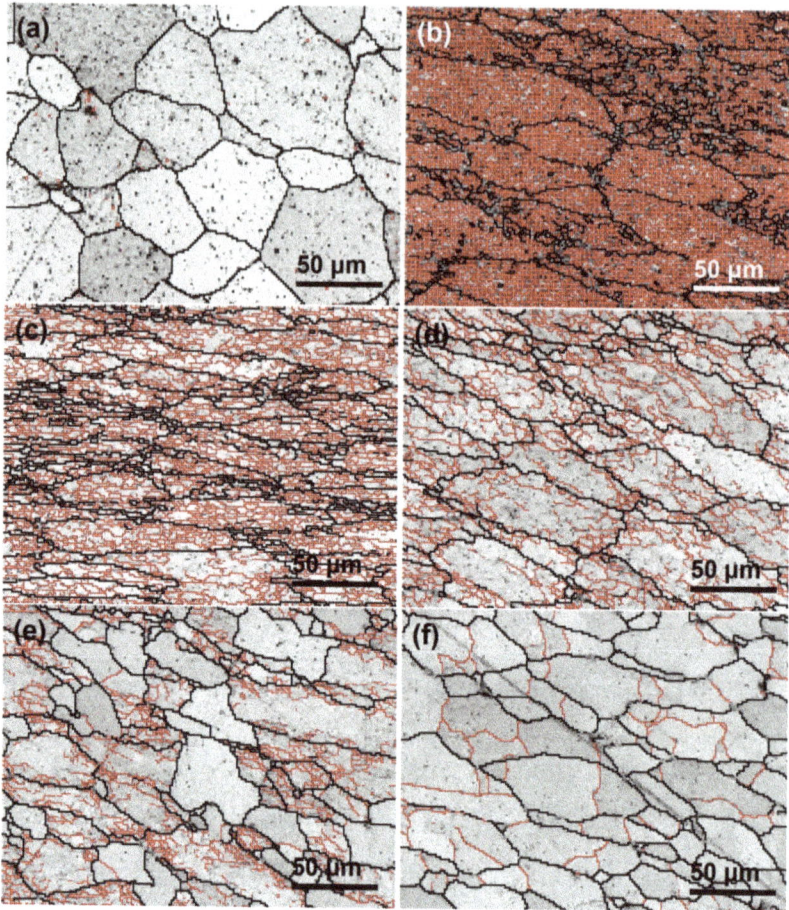

Figure 5. EBSD images of grain morphology of spray-deposited 7055 aluminum alloy (**a**) and under hot compression conditions of 300 °C/5 s^{-1} (**b**), 300 °C/0.001 s^{-1} (**c**), 400 °C/0.1 s^{-1} (**d**), 450 °C/5 s^{-1} (**e**) and 450 °C/0.001 s^{-1} (**f**) (Black line represents the high angle grain boundaries (HAGBs) which are higher than 15°, and the red line represents the low angle grain boundaries (LAGBs) ranging from 2° to 15°).

The Zener–Hollomon (Z) parameter is described as the temperature-compensated strain rate factor. The value of Z decreases with the decrease in the strain rate or the increase in the deformation temperature. With the decrease in the Z value, the DRX degree, the DRX grain size and the subgrain size all increase gradually, but the substructure volume fraction decreases. When compressed at 450 °C, the equiaxed DRX grains replace the original ones (Figure 5e,f). It can be concluded that the spray-deposited 7055 aluminum alloy undergoes almost complete DRX during compression at 450 °C. Due to the sufficient growth of the substructure under high-temperature conditions, the DRX nucleation should be dominated by the progressive transformation of subgrains. The DRX grains also grow sufficiently under low strain rates, as shown in Figure 5f.

3.2.3. The Substructures

Figure 6 shows the substructures of the as-deformed specimens of the 7055 aluminum alloy. In addition to the micron-level second phase, there are also amounts of smaller second phases with the size of 200~300 nm. At 300 °C, the density of the second phase and deformation substructure is highest in the specimen deformed with $5s^{-1}$, and the substructure size is the smallest (ranging from 0.5 to 2 µm). It is worth noting that the average size of the subgrains is almost the same as that of the residual coarse phase, as shown in Figure 6a. With the decrease in the Z parameter, the subgrain size increases. The average size of the subgrains ranges from 5 to 8 µm in the specimen deformed at 300 °C with $0.001\ s^{-1}$. The residual coarse second phase distributes on the subgrain boundary (Figure 6b). Low density dislocation entanglement also exists in subgrains.

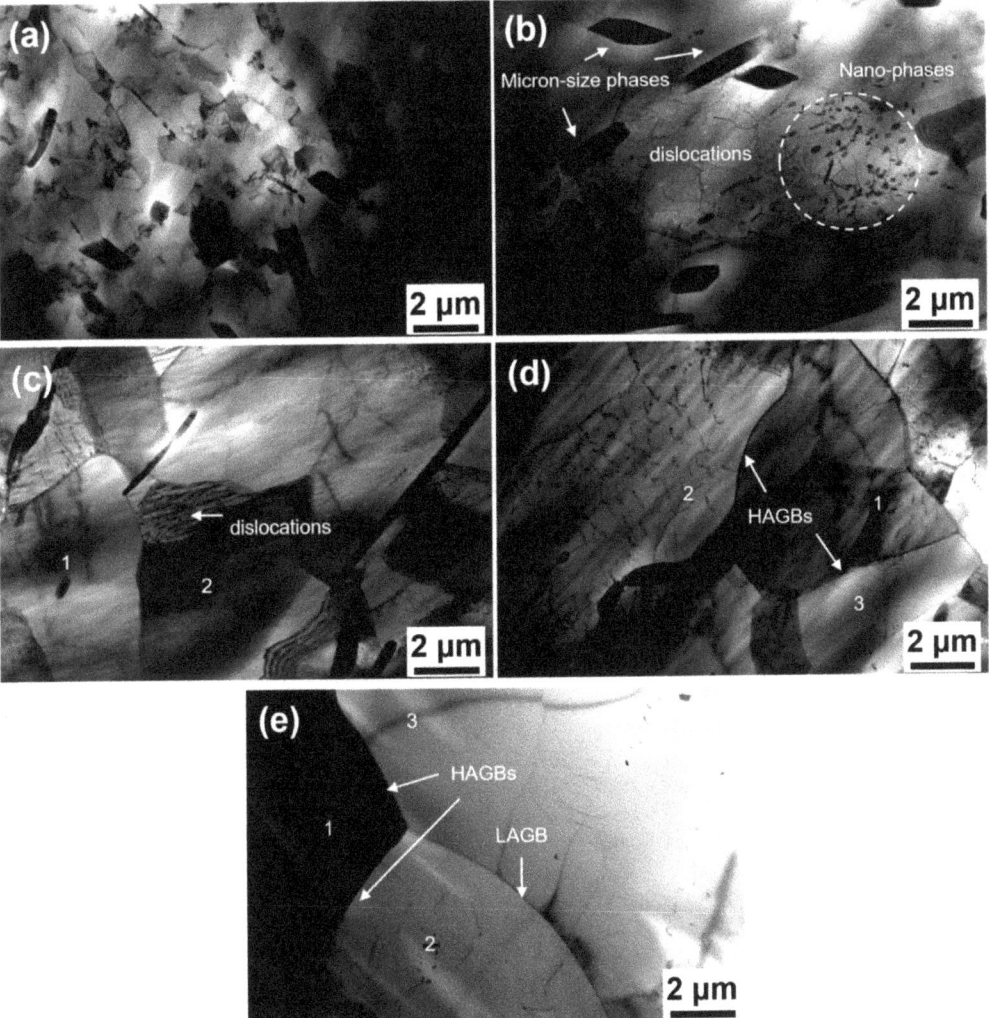

Figure 6. TEM images of the sub-microstructure of spray-deposited 7055 aluminum alloy under hot compression conditions of 300 °C/5 s^{-1} (**a**), 300 °C/0.001 s^{-1} (**b**), 400 °C/0.1 s^{-1} (**c**), 450 °C/5 s^{-1} (**d**) and 450 °C/0.001 s^{-1} (**e**).

At 400 °C/1 s^{-1} and 450 °C/5 s^{-1} deformation conditions, only a very small amount of the micron second phase remains (Figure 6c). The subgrains develop maturely and some dislocation-free subgrains can be observed (Figure 6d). In some grains, dislocation entanglement and cellular structure can also be observed. The dynamic softening behavior consumes deformation energy storage through dislocation slip and climbing, and even results in DRX. However, the dynamic hardening behavior constantly introduces dislocations, and the new dislocations cannot be consumed by the growth of substructures in time. Therefore, sufficient subgrain growth and dislocation entanglement coexists in thermally deformed samples. In addition, large-size subgrains 1 and 2 or 3 have an obvious contrast under the same incident condition (Figure 6d), which indicates that the grain boundary between grains 1 and 2 or 1 and 3 has large misorientation. That means grain 1 was a recrystallized nucleus or will grow into the nucleus if the deformation continues.

Under the 450 °C/0.001 s^{-1} deformation condition, the second phase of the micron size is completely dissolved. As depicted in Figure 6e, the misorientation between large subgrains 1 and 2 or 1 and 3 was more prominent, which indicates that the flat interface between grains 1 and 2 or 1 and 3 should also be HAGB. The large subgrain 1 developed into a DRX nucleus. Besides, the subgrain boundary between grains 2 and 3 shows a large curvature. It can be inferred that grains 2 and 3 will eventually develop into another DRX nucleus by subgrain boundary migration. In addition, at 450 °C, few residual dislocations were observed in the 0.001 s^{-1} deformed specimen. Unlike the high-speed deformed samples at 450 °C, the strain rate of 0.001 s^{-1} is very low, so the new dislocations have enough time to be consumed by substructure growth or recrystallization nucleation (Figure 6e).

4. Discussion

As we all know, the fault energy γ_{SFE} determines the extent to which unit dislocations dissociate into partial dislocations. For materials with high γ_{SFE}, such as aluminum (166 mJ·m^{-2}), the dissociating of the dislocation into two partials is more difficult. Therefore, it is generally believed that Al-Zn-Mg-Cu alloys mainly undergo DRV during hot deformation, and only partial DRX occurs under the condition of a low Z value [25]. Because most of the DRX grains appear adjacent to the original HAGBs, the primary DRX nucleation theories of the 7055 aluminum alloy are "strain-induced HAGB migration (SIBM)" and "subgrain rotation near HAGBs". It should be noted that the content of alloying elements of the high-strength 7000 series aluminum alloy can exceed 20% (mass fraction). The fault energies of Zinc, Magnesium and Copper are 140, 74~125 and 78 mJ·m^{-2} [25,26], respectively. Studies [26,27] show that the co-addition of these alloying elements can effectively reduce the fault energy of the aluminum alloy, and then affect the movement behavior and the dislocation configuration, resulting in a dynamic softening mechanism different from that of pure aluminum. In addition, the interaction between a large number of micron-scale second phases and dislocations will also change the configuration of dislocations, thus enriching the dynamic softening behavior.

4.1. Interaction between the Second Phase and Dislocation

The residual coarse phases ranging from 1 to 2 μm under different Z parameters are depicted in Figure 7. It can be seen from Figure 7a that there are a lot of fine sub-structures and cellular structures when deformed at 300 °C with 5 s^{-1}. This typical DRV behavior can be explained as follows: on the one hand, there is no time and a lack of thermal activation to obtain an adequate annihilation or rearrangement degree of dislocations. On the other hand, due to the pile-up of dislocations in front of the large second phase, a rapid formation of sub-microstructures can be stimulated by the dislocation packing. A different contrast indicates a large misorientation between subgrains 1 and 2. The right side of subgrain 1 is close to the residual coarse second phase. The high-density dislocation configuration in front of the second phase will produce a subgrain with relatively large misorientation.

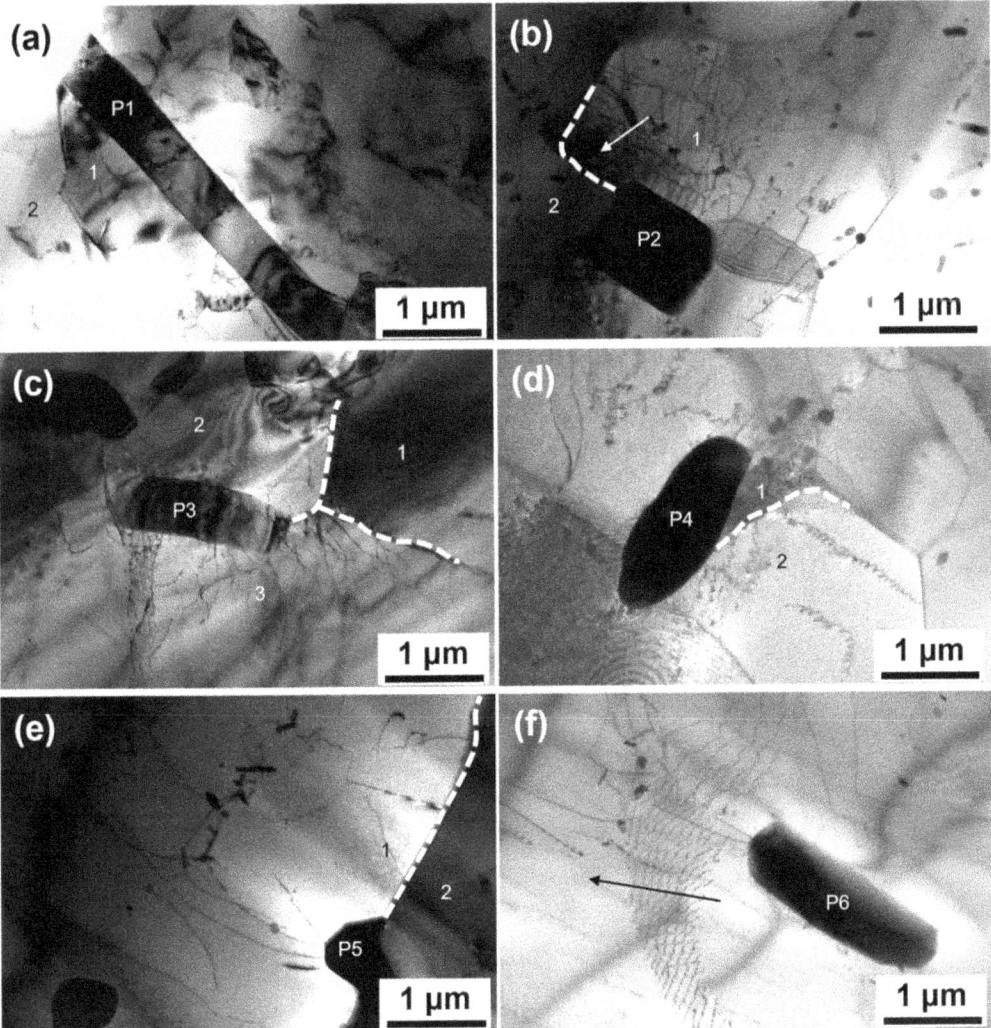

Figure 7. TEM images of interaction between dislocation and second phase of spray-deposited 7055 aluminum alloy under hot compression conditions of 300 °C/5 s^{-1} (a), 300 °C/0.001 s^{-1} (b), 400 °C/0.1 s^{-1} (c), 450 °C/5 s^{-1} (d) and 450 °C/0.001 s^{-1} (e,f); P$_i$ (i = 1~6) represents the particles that remained after hot compressed.

In Figure 7b, the dislocation entanglement can be observed at the front of the second phase (the area pointed by the arrow). The local interface (dotted line) between subgrains 1 and 2 continuously absorbs the stacking dislocations, resulting in a misorientation increase. When the subgrain boundary evolves to a HAGB, DRX nucleation completes. In Figure 7c, there is a particle P3 between subgrains 2 and 3. Subgrain 1 has no contact with P3. Therefore, the misorientation between subgrains 2 and 3 is discernible, while that between subgrains 2/3 and 1 is relatively larger. In Figure 7d, the dislocation density decreases significantly due to the high compression temperature. The grain boundary of subgrain 2 continuously absorbs the lattice dislocations near the P4 to increase its misorientation, and finally evolves into a HAGB. Figure 7e,f show the second phase morphology on the grain boundary and inside the grain of the specimen deformed at 450 °C with 0.001 s^{-1}, respectively. The residual second phase inner grain serves as the source of dislocation generation, providing dislocations to the grain boundary (the arrow represents the direction of the dislocation slip). At this time, the climbing ability of dislocation is enhanced. There is enough time to finish polygonization and annihilation. Therefore, the dislocation stacking at the front of the second phase or the grain boundary is insignificant anymore.

It can be concluded that, under high Z value conditions, the volume fraction of the residual second phase increases. The dislocation climbing and cross slip capabilities are restricted. Under this condition, the interaction behaviors between the second phase and dislocations consist of "dislocation entanglement" and "dislocation-packing induced subgrains forming" in front of the second phase. There is a relatively large misorientation between the subgrains near the second phase and those faraway. At low Z values, the volume fraction of the residual second phase decreases. Dislocations are thermally activated without being effectively blocked.

4.2. Dynamic Recovery Behavior

It can be seen from Figure 8 that the substructure size increases with the decrease in Z. At 300 °C, there are dislocation cells (<1μm) in the specimen deformed with 5 s^{-1}. The left side of the cell boundary is composed of the dislocation wall (the white dotted line), while the right side retains a high-energy dislocation network. Under the high strain rate and low temperature conditions, it is a lack of time for the dislocation migrating to the cell boundary. The 300 °C/0.001 s^{-1} deformed specimen also has the dislocation wall. The walls form a "Y" shape, which divides one substructure into three parts. However, the dislocation network cannot be observed on segmented intracellular regions. Compared with the deformed sample at 300 °C with 5 s^{-1}, the dislocations have enough time to construct a low-energy configuration. The remaining scattered dislocations inside the cell will also continue to slip into the dislocation wall. Annihilation of unlike dislocations and the rearrangement of dislocations coexist. Thus, the density of dislocations with the same type increases in the dislocation wall. The grain boundary misorientation increases continuously.

At 400 °C/0.1 s^{-1} and 450 °C/5 s^{-1} conditions, the substructures with the size of 1~2 μm are dislocation-free. Part of the boundary of this substructure is flat and sharp (as shown by the white dashed line in Figure 8c,d) and has evident misorientation to the adjacent microstructure. Another part of the interface of this structure is characterized by low-density interfacial dislocations. These mobile dislocations are consuming the adjacent dislocation network through interfacial migration, as shown by the arrows.

It can be concluded that this kind of substructure has almost finished the transformation to the subgrain. The subgrains 1 and 2 in Figure 8c,d will grow sufficiently by the local interface migration. As shown in Figure 8e, the subgrains in the sample deformed at 450 °C/0.001 s^{-1} grew completely, and their size can reach a scale of tens of microns. The few remaining dislocations in the subgrain arrange to a low energy state and tend to merge into the subgrain boundary.

4.3. Dynamic Recrystallization Behavior

Dynamic recrystallization requires higher thermal activation. EBSD analysis reveals that there is no apparent DRX in the 300 °C/5 s^{-1} deformed specimen. DRX nuclei of specimens deformed at 300 °C/0.001 s^{-1}, 400 °C/0.1 s^{-1}, 450 °C/5 s^{-1} and 450 °C/0.001 s^{-1} are shown in Figure 9. The size of the recrystallized nucleus ranging from 8 to 10 μm is larger than that of the DRV substructure. In the 300 °C/0.001 s^{-1} condition, the dislocation-free subgrains containing a coarse second phase can be observed (as shown in the solid line frame in Figure 9a). Similar features also appear in specimens deformed at 400 °C/0.1 s^{-1} (solid line frame in Figure 9b). Research shows that the particle-stimulated nucleation (PSN) occurs during the static recrystallization of cold-deformed metals [27]. Similarly, under the condition of a high temperature and large strain (the compression in this experiment is 60%), the DRX based on PSN can also be observed. Zang [28] et al. studied the hot deformation behavior of the Al-7.9 Zn-2.7 Mg-2.0 Cu alloy sheet during the hot rolling process. The results show that the residual coarse second phase after homogenized heat treatment can also produce particle excitation nucleation.

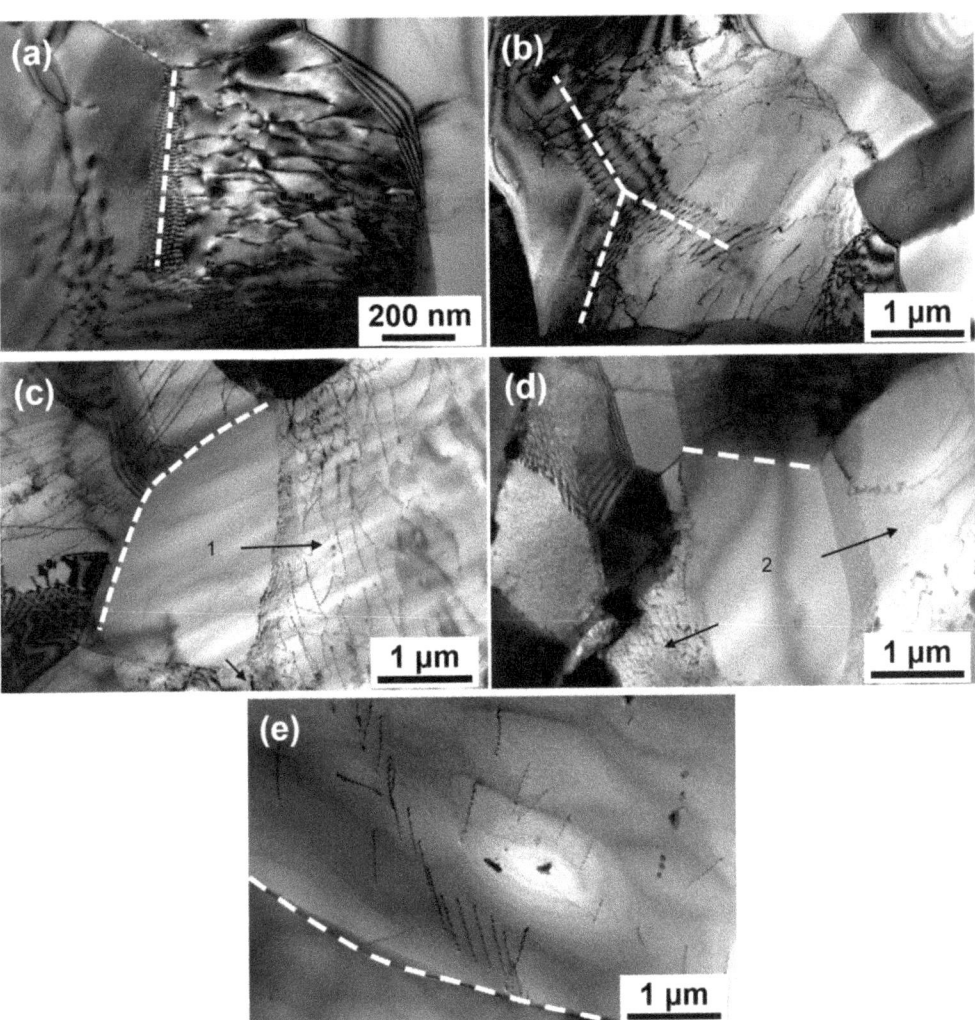

Figure 8. TEM images of DRV microstructure of spray-deposited 7055 aluminum alloy under hot compression conditions of 300 °C/5 s^{-1} (**a**), 300 °C/0.001 s^{-1} (**b**), 400 °C/0.1 s^{-1} (**c**), 450 °C/5 s^{-1} (**d**) and 450 °C/0.001 s^{-1} (**e**).

Subgrains with a large curvature and HAGBs can be observed in the 400 °C/0.1 s^{-1} and 450 °C/5 s^{-1} deformed specimens. The local grain boundaries of the subgrains are bulging out to the high-density dislocation (depicted by the arc-shaped grain boundaries and migration arrow directions in Figure 9b,c). The dislocation network will be swept by the bulging of the subgrain boundary. Then, the misorientation increases continuously, eventually forming the recrystallized nucleus. It indicates that subgrain boundary migration is also the DRX nucleation mechanism of the spray-deposited 7055 aluminum alloy. Under the deformation condition of 450 °C/0.001 s^{-1}, the polygonal recrystallized nuclei with flat grain boundaries appeared (the grains 1 and 2 shown in Figure 9d). There is also a small subgrain at the grain boundary triple junctions between grains 1, 2 and 3. Under the migration of grain boundaries of grains 1 and 2, the small subgrain will disappear. It is a typical characteristic of the nucleation mechanism named "subgrain boundary migration".

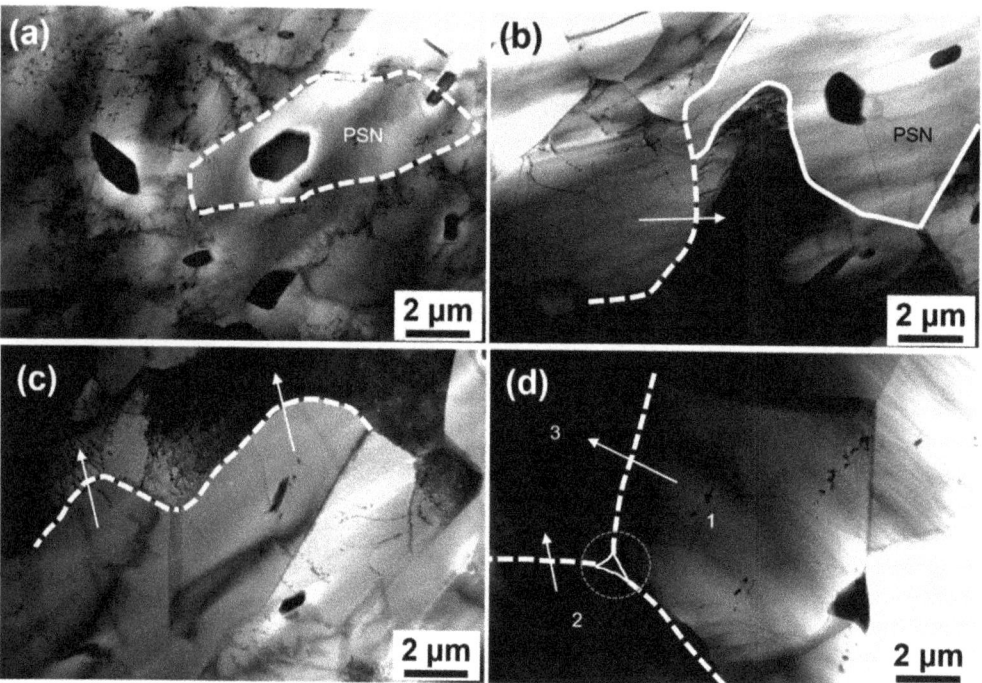

Figure 9. TEM images of DRX microstructure of spray-deposited 7055 aluminum alloy under hot compression conditions of 300 °C/0.001 s^{-1} (a), 400 °C/0.1 s^{-1} (b), 450 °C/5 s^{-1} (c) and 450 °C/0.001 s^{-1} (d).

Figure 10 shows the inverse pole figure (IPF) diagrams of the spray-deposited 7055 aluminum alloy under different hot compression conditions. Figure 11 shows the statistics of cumulative misorientation of subgrains along the arrows shown in Figure 10. Under different Z parameters, the accumulative misorientations from the interior grain to the HAGBs all indicate a significant increase in misorientation (12~20 °). Research shows that "subgrain rotation near original grain boundary" occurs when the cumulative misorientation in the grain exceeds 10°. The so-called subgrain rotation mechanism can be described as follows: There is a small misorientation between two adjacent subgrains. The dislocation network on the subgrain boundary can be dissociated, disassembles and transfers to the adjacent subgrain boundary, resulting in the disappearance of the subgrain boundary, and then a nucleus formed. Since favorably slip systems are always activated adjacent to the HAGBs, it is easier to create a recrystallized nucleus when the subgrain rotation occurs at the front of the original HAGBs. This also explains why dynamic recrystallization occurs preferentially at the original grain boundary, as shown in Figure 10a. With the decrease in the Z parameter, the recrystallized nucleus grows up and nucleates again at the newly formed HAGB. Therefore, the DRX characteristics as a necklace at the original grain boundary become less evident with the decrease in the Z parameter, as shown in Figure 10b–d.

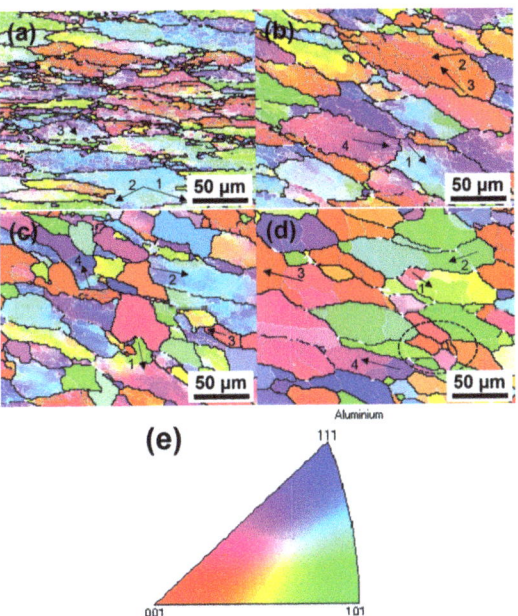

Figure 10. IPF images of the deformed microstructure of spray deposited 7055 aluminum alloy under hot compression conditions of 300 °C/0.001 s^{-1} (**a**), 400 °C/0.1 s^{-1} (**b**), 450 °C/5 s^{-1} (**c**) and 450 °C/0.001 s^{-1} (**d**). (**e**) Representation of the color code used to identify the crystallographic orientation on a standard stereographic (Arrows 1, 2, 3 and 4 in each image are the misorientation cumulative routes).

Besides, at relatively high temperatures, a homogeneous microstructure usually develops (Figure 5e,f). Research reported that the CDRX occurs by the progressive accumulation of dislocations into the low angle grain boundaries (LAGBs) which increase the misorientation. Eventually, HAGBs are formed when the misorientation reaches a critical value of about 15°.

As shown in Figure 10d, several fine recrystallized grains gather together (in the dashed box). This characteristic is not only different from the result after the subgrain rotation but also different from the result of the subgrain boundary migration, indicating a progressive misorientation increase in subgrains.

Figure 12 is a diagrammatic sketch of microstructure evolution. The microstructure evolution of the spray-deposited 7055 aluminum alloy can be divided into three cases: (1) DRV behavior at a low temperature and high strain rate. The second phase is less dissolved. High-density and small-size subgrains appear; (2) DRV and DRX behaviors at an intermediate strain rate and temperature. The amount of the second phase dissolved increases with the increase in the deformation temperature. However, there is still a small amount of an undissolved and large-sized second phase, resulting in particle stimulated nucleation (PSN). There is also subgrain boundary migration nucleation inside the grains. However, the primary recrystallization nucleation mechanism is the subgrain rotation at the HAGBs; (3) Almost complete dynamic recrystallization occurs at high temperatures and low strain rates. In this condition, the subgrains are well defined first. Recrystallization nuclei are formed by the mechanism of the homogeneous misorientation increase, which can be considered as continuous dynamic recrystallization.

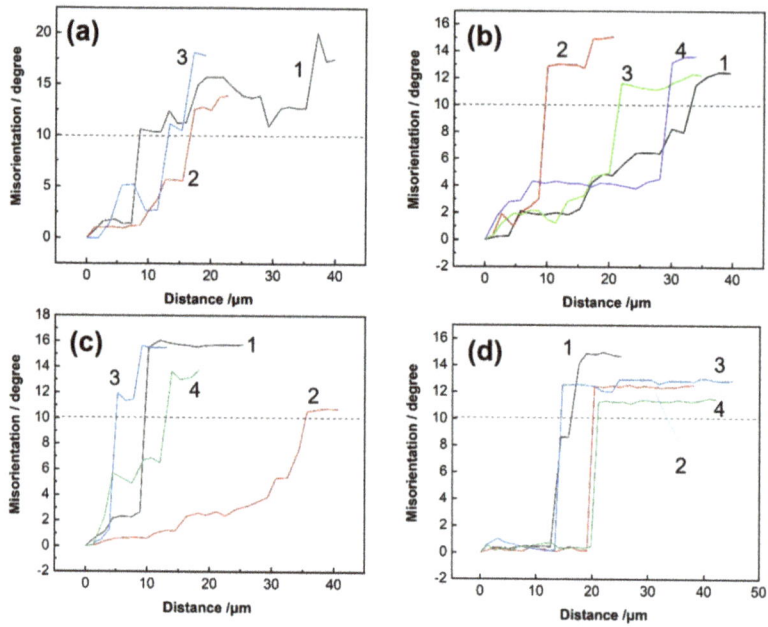

Figure 11. Misorientation cumulative images between deformed subgrains of spray-deposited 7055 aluminum alloy under hot compression conditions of 300 °C/0.001 s^{-1} (a), 400 °C/0.1 s^{-1} (b), 450 °C/5 s^{-1} (c) and 450 °C/0.001 s^{-1} (d). (Curves 1, 2, 3 and 4 in each image are the misorientation cumulative routes in Figure 10).

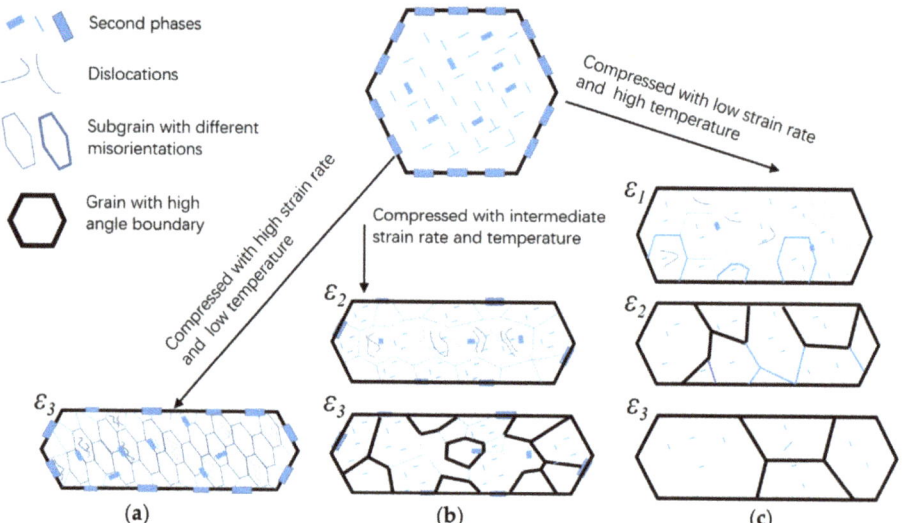

Figure 12. Diagrammatic sketch of the microstructure evolution of spray-deposited 7055 aluminum alloy. (**a**) Dynamic recovery and a few second phases resolve. (**b**) Dynamic recrystallization based on the "subgrain rotation", "subgrain boundary migration" and "PSN" nucleated mechanism. (**c**) Dynamic recrystallization based on "homogeneous misorientation increase of subgrain", and a few phases exist.

When compared with the dynamic recrystallization behavior of the 7055 aluminum alloy produced by semi-continuous casting [16,17], the spray-deposited 7055 aluminum alloy showed the complex DRX mechanisms and a higher DRX degree, which are beneficial to reduce the deformation resistance and improve the deformation ability.

5. Conclusions

(1). The AlZnMgCu phase in the spray-deposited 7055 aluminum alloy gradually dissolves with the increase in the deformation temperature, while the Al_7Cu_2Fe phase does not change. The residual AlZnMgCu phase can induce the rapid formation of subgrains and produce particle stimulated nucleation (PSN) recrystallization.

(2). The plastic instability of the spray-deposited 7055 aluminum alloy occurs at 470 °C with 1~5 s^{-1} strain rates. DRV and DRX occur under other strain conditions simultaneously. The DRX behavior is evident at low Z parameters.

(3). The DRX nucleation mechanism at 300~400 °C and 0.001~0.1 s^{-1} is "subgrain rotation near the original HAGBs" and "subgrain boundary migration". Under a 450 °C deformation temperature with a low strain rate, the nucleation mechanism is considered "the homogeneous misorientation increase of subgrain". At the high strain rate of 300~400 °C, "residual coarse second phase particle stimulated nucleation" also occurs.

Author Contributions: Investigation, Methodology, Data Curation, Writing—original draft, D.F. and R.X.; Resources, Conceptualization, Formal analysis, Writing—review and editing, J.L. and W.H.; Conceptualization, Data curation, C.L. and Y.L.; Funding Acquisition, H.Z.; Validation, Supervision, J.W. and L.Z. All authors have read and agreed to the published version of the manuscript.

Funding: This research was funded by the National Natural Science Foundation of China (grant No.52205421), the Key Research and Development Program of Zhenjiang City (grant Nos. GY2021003 and GY2021020), Graduate Research and Innovation Projects in Jiangsu Province (grant No.KYCX21_3453) and Undergraduate Innovation and Entrepreneurship Training Program of Jiangsu Province (grant No.202210289112Y).

Institutional Review Board Statement: Not applicable.

Informed Consent Statement: Not applicable.

Data Availability Statement: Not applicable.

Acknowledgments: The authors would like to appreciate the financial supports from the National Natural Science Foundation of China (grant No.52205421), the Key Research and Development Program of Zhenjiang City (grant Nos. GY2021003 and GY2021020), the Graduate Research and Innovation Projects in Jiangsu Province (grant No.KYCX21_3453) and the Undergraduate Innovation and Entrepreneurship Training Program of Jiangsu Province (grant No.202210289112Y).

Conflicts of Interest: The authors declare no conflict of interest.

References

1. Watzl, G.; Grünsteidl, C.; Arnoldt, A.; Nietsch, J.A.; Österreicher, J.A. In situ laser-ultrasonic monitoring of elastic parameters during natural aging in an Al-Zn-Mg-Cu alloy (AA7075) sheet. *Materialia* **2022**, *26*, 101600. [CrossRef]
2. Luo, R.; Cao, Y.; Bian, H.K.; Chen, L.L.; Peng, C.T.; Cao, F.Y.; Ouyang, L.X.; Qiu, Y.; Xu, Y.J.; Chiba, A.; et al. Hot workability and dynamic recrystallization behavior of a spray formed 7055 aluminum alloy. *Mater. Charact.* **2021**, *178*, 111203. [CrossRef]
3. Li, J.C.; Feng, D.; Xia, W.S.; Guo, W.M.; Wang, G.Y. The Non-Isothermal Double Ageing Behaviour of 7055 Aluminum Alloy. *Acta Metall. Sin.* **2020**, *56*, 1496–1508. Available online: https://www.ams.org.cn/CN/10.11900/0412.1961.2020.00039 (accessed on 15 November 2022).
4. Schreiber, J.M.; Omcikus, Z.R.; Eden, T.J.; Sharma, M.M.; Champagne, V.; Patankar, S.N. Combined effect of hot extrusion and heat treatment on the mechanical behavior of 7055 AA processed via spray metal forming. *J. Alloys. Compd.* **2014**, *617*, 135–139. [CrossRef]
5. Wu, C.H.; Feng, D.; Ren, J.J.; Zang, Q.H.; Li, J.C.; Liu, S.D.; Zhang, X.M. Effect of non-isothermal retrogression and re-ageing on through-thickness homogeneity of microstructure and properties of Al-8Zn-2Mg-2Cu alloy thick plate. *J. Cent. South Univ.* **2022**, *29*, 960–972. [CrossRef]
6. Tang, J.; Zhan, H.; Hang, H.; Teng, J.; Fu, D.F.; Jiang, F.L. Effect of Zn content on the static softening behavior and kinetics of Al–Zn–Mg–Cu alloys during double-stage hot deformation. *J. Alloys Compd.* **2019**, *806*, 1081–1096. [CrossRef]

7. Wu, C.H.; Feng, D.; Zang, Q.H.; An, S.C.; Zhang, H.; Lee, Y.S. Microstructure Evolution and Recrystallization Behavior During Hot Deformation of Spray Formed AlSiCuMg Alloy. *Acta Metall. Sin.* **2022**, *58*, 932–943. Available online: https://www.ams.org.cn/CN/10.11900/0412.1961.2021.00329 (accessed on 15 November 2022).
8. Feng, D.; Zhu, T.; Zang, Q.H.; Lee, Y.S.; Fan, X.; Zhang, H. Solution Behavior of Spray-Formed Hypereutectic AlSiCuMg Alloy. *Acta Metall. Sin.* **2022**, *58*, 1129–1141. Available online: https://www.ams.org.cn/CN/10.11900/0412.1961.2021.00079 (accessed on 15 November 2022).
9. Khan, M.A.; Wang, Y.W.; Anjum, M.J.; Yasin, G.; Malik, A.; Nazeer, F.; Khan, S.; Ahmad, T.; Zhang, H. Effect of heat treatment on the precipitate behaviour, corrosion resistance and high temperature tensile properties of 7055 aluminum alloy synthesis by novel spray deposited followed by hot extrusion. *Vacuum* **2020**, *174*, 109185. [CrossRef]
10. Liu, L.L.; Pan, Q.L.; Wang, X.D.; Xiong, S.W. The effects of aging treatments on mechanical property and corrosion behavior of spray formed 7055 aluminum alloy. *J. Alloys Compd.* **2018**, *735*, 261–276. [CrossRef]
11. Lin, X.M.; Cao, L.F.; Wu, X.D.; Tang, S.B.; Zou, Y. Precipitation behavior of spray-formed aluminum alloy 7055 during high temperature aging. *Mater. Charact.* **2022**, *9*, 112347. [CrossRef]
12. Jiang, Y.M.; Zhao, Y.; Zhao, Z.X.; Yan, K.; Ren, L.T.; Du, C.Z. The strengthening mechanism of FSWed spray formed 7055 aluminum alloy under water mist cooling condition. *Mater. Charact.* **2020**, *162*, 110185. [CrossRef]
13. Ma, S.C.; Zhao, Y.; Pu, J.H.; Zhao, Z.X.; Liu, C.; Yan, K. Effect of welding speed on performance of friction stir welded spray forming 7055 aluminum alloy. *J. Manuf. Process.* **2019**, *46*, 304–316. [CrossRef]
14. Huang, T.; Xun, J.H.; Yun, L.H.; Cheng, Y.; Hua, Y.X.; Zhang, H. Study on ductile fracture of unweldable spray formed 7055 aluminum alloy TIG welded joints with ceramic particles. *Mater. Today Comm.* **2021**, *29*, 102835. [CrossRef]
15. Feng, D.; Zhang, X.M.; Liu, S.D.; Deng, Y.L. Constitutive equation and hot deformation behavior of homogenized Al-7.68Zn-2.12Mg-1.98Cu-0.12Zr alloy during compression at elevated temperature, *Mater. Sci. Eng. A* **2014**, *608*, 63–72. [CrossRef]
16. Feng, D.; Wang, G.Y.; Chen, H.M.; Zhang, X.M. Effect of Grain Size Inhomogeneity of Ingot on Dynamic Softening Behavior and Processing Map of Al-8Zn-2Mg-2Cu alloy. *Met. Mater. Int.* **2018**, *24*, 195–204. [CrossRef]
17. Feng, D.; Zhang, X.M.; Liu, S.D.; Han, N.M. Effect of Grain Size on Hot Deformation Behavior of a New High Strength Aluminum Alloy. *Rare Met. Mater. Eng.* **2018**, *45*, 2014–2021. Available online: https://www.engineeringvillage.com/app/doc/?docid=cpx_7e3821a415743cb921fM6d3710178163171&pageSize=25&index=1&searchId=dd7704772cfb47c1bd6df21ebb694e86&resultsCount=20&usageZone=resultslist&usageOrigin=searchresults&searchType=Quick (accessed on 15 November 2022).
18. Liu, S.D.; Wang, S.L.; Ye, L.Y.; Deng, Y.L.; Zhang, X.M. Flow behavior and microstructure evolution of 7055 aluminum alloy impacted at high strain rates. *Mater. Sci. Eng. A* **2016**, *677*, 203–210. [CrossRef]
19. Yang, Q.Y.; Deng, Z.H.; Zhang, Z.Q.; Liu, Q.; Jia, Z.H.; Huang, G.J. Effects of strain rate on flow stress behavior and dynamic recrystallization mechanism of Al-Zn-Mg-Cu aluminum alloy during hot deformation. *Mater. Sci. Eng. A* **2016**, *662*, 204–213. [CrossRef]
20. Zang, Q.H.; Yu, H.S.; Lee, Y.S.; Kim, M.S.; Kim, H.W. Effects of initial microstructure on hot deformation behavior of Al-7.9Zn-2.7Mg-2.0Cu (wt%) alloy. *Mater. Charact.* **2019**, *151*, 404–413. [CrossRef]
21. Yu, H.C.; Wang, M.P.; Sheng, X.F.; Li, Z.; Chen, L.B.; Lei, Q.; Chen, C.; Jia, Y.L.; Xiao, Z.; Chen, W.; et al. Microstructure and tensile properties of large-size 7055 aluminum billets fabricated by spray forming rapid solidification technology. *J. Alloys Comp.* **2013**, *578*, 208–214. [CrossRef]
22. Feng, D.; Han, Z.J.; Li, J.C.; Zhang, H.; Xia, W.S.; Fan, X.; Tang, Z.H. Evolution Behavior of Primary Phase During Pre-heat Treatment Before Deformation for Spray Formed 7055 Aluminum Alloy. *Rare Met. Mat. Eng.* **2020**, *49*, 4253–4264. Available online: https://www.engineeringvillage.com/app/doc/?docid=cpx_5d7fe0a117788cc7620M6f6910178163190&pageSize=25&index=1&searchId=340b1898f4c045e99932ad7af2342fba&resultsCount=1&usageZone=resultslist&usageOrigin=searchresults&searchType=Quick (accessed on 15 November 2022).
23. Xiang, K.Y.; Ding, L.P.; Jia, Z.H.; Xie, Z.Q.; Fan, X.; Ma, W.T.; Zhang, H. Research progress of ultra-high strength spray-forming Al-Zn-Mg-Cu alloy. *Chin. J. Nonferrous Met.* **2022**, *32*, 1199–1224. [CrossRef]
24. Xie, Z.Q.; Jia, Z.H.; Xiang, K.Y.; Kong, Y.P.; Li, Z.G.; Fan, X.; Ma, W.T.; Zhang, H.; Lin, L.; Marthinsen, K.; et al. Microstructure evolution and recrystallization resistance of a 7055 alloy fabricated by spray forming technology and by conventional ingot metallurgy. *Metall. Mater. Trans. A* **2020**, *51*, 5378–5388. [CrossRef]
25. LI, J.C.; Wu, X.D.; Liao, B.; Lin, X.M.; Cao, L.F. Simulation of low proportion of dynamic recrystallization in 7055 aluminum alloy. *Trans. Nonferrous Met. Soc. China* **2021**, *31*, 1902–1915. [CrossRef]
26. Huang, K.; Logé, R.E. A review of dynamic recrystallization phenomena in metallic materials. *Mater. Des.* **2016**, *111*, 548–574. [CrossRef]
27. Shao, Y.; Shi, J.H.; Pan, J.C.; Liu, Q.H.; Yan, L.; Guo, P.Y. Influence of thermo-mechanical conditions on the microstructure and mechanical property of spray-formed 7055 aluminum alloy. *Mater. Today Comm.* **2022**, *31*, 103593. [CrossRef]
28. Zang, Q.H.; Chen, H.M.; Lee, Y.S.; Yu, H.S.; Kim, M.S.; Kim, H.W. Improvement of anisotropic tensile properties of Al-7.9Zn-2.7Mg-2.0Cu alloy sheets by particle stimulated nucleation. *J. Alloys Compd.* **2020**, *828*, 154330. [CrossRef]

Article

Comparative Study and Multi-Objective Crashworthiness Optimization Design of Foam and Honeycomb-Filled Novel Aluminum Thin-Walled Tubes

Yi Tao [1], Yonghui Wang [2], Qiang He [2,*], Daoming Xu [1] and Lizheng Li [2]

1 Marine Design & Research Institute of China, Shanghai 200011, China
2 School of Mechanical Engineering, Jiangsu University of Science and Technology, Zhenjiang 212000, China
* Correspondence: heqiang@just.edu.cn

Abstract: Due to their lightweight, porous and excellent energy absorption characteristics, foam and honeycomb materials have been widely used for filling energy absorbing devices. For further improving the energy absorption performance of the novel tube proposed in our recent work, the nonlinear dynamics software Abaqus was firstly used to establish and verify the simulation model of aluminum-filled tube. Then, the crashworthiness of honeycomb-filled tubes, foam-filled tubes and empty tube under axial load was systematically compared and analyzed. Furthermore, a comparative analysis of the mechanical behavior of filled tubes subjected to bending load was carried out based on the study of dynamic response curve, specific energy absorption and deformation mechanism, the difference in energy absorption performance between them was also revealed. Finally, the most promising filling structure with excellent crashworthiness under lateral load was optimized. The research results show that the novel thin-walled structures filled with foam or honeycomb both show better energy absorption characteristics, with an increase of at least 8.8% in total absorbed energy. At the same time, the mechanical properties of this kind of filled structure are closely related to the filling styles. Foam filling will greatly damage the weight efficiency of the novel thin-walled tube. However, honeycomb filling is beneficial to the improvement of SEA, which can be improved by up to 18.2%.

Keywords: thin-walled structure; filling structure; numerical simulation; crashworthiness optimization; energy absorption performance

1. Introduction

Thin-walled structures have been widely used in automobiles, aviation and other industrial fields, due to their excellent mechanical properties. Therefore, the research on their mechanical properties has always been a hot topic for scholars [1–5]. Galib et al. [6] conducted a comprehensive experimental and numerical study on circular tubes subjected to dynamic load. Zhang et al. [7,8] pointed out that multi-cell square tubes showed better mechanical properties. Alavi Nia et al. [9] conducted axial impact tests on structures with different polygonal cross-sections, and proposed that the multi-cell cross-section was conducive to the improvement of energy absorption performance.

Considering that the traditional thin-walled structure has limited room to improve energy absorption efficiency and stability, it can no longer meet current requirements. Researchers have found that applying biological structural features to structural design can effectively enhance its energy absorption performance [10–13]. Song et al. [14] designed a novel bionic tube with grooves and studied its crashworthiness under lateral impact. The study demonstrated that the innovative design is conducive to improving the energy absorption efficiency of regular structures. Based on the structural characteristics of bamboo, Zou et al. [15] designed a bionic tube and solved its numerical examples under axial/transverse impact. Palombini et al. [16] mechanically explored the special geometry of a single vascular bundle in bamboo, and the new design proposed has a better improvement in its strength and crashworthiness under dynamic loads. Ferdynus et al. [17] proposed a new type of trigger for the square tube and focused on its effect on the energy absorption indicators achieved (triggering effect).

Metal matrix syntactic foams are high-performance foams consisting of a light-weight matrix and a set of porous fillers. Orbulov and Szlancsik et al. [18–20] carried out a lot of experimental work to characterize its structure–mechanical property relationship. Fiedler et al. [21] analyzed the mechanical properties of the foam with gradient characteristics. Rabiei et al. [22,23] manufactured steel composite metal foam core sandwich panels and studied their quasi-static mechanical properties. These studies show that metal foam has superior mechanical properties and can be used as energy absorption materials.

In order to further enhance the crashworthiness, some scholars fill the regular tubes with lightweight porous materials such as metal foam and honeycomb [24–28]. Li et al. [29,30] conducted bending experiments on foam-filled tubes with different structures. Qi et al. [31] conducted a numerical analyzed mechanical behavior investigation of empty and foam-filled hybrid beams, and optimized their design. Pandarkar et al. [32] elaborated on foam-filled pipes, and the main conclusion was that filling thin-walled structures can improve the stability of the structure. Cakıroglu [33] focused on the quasi-static mechanical properties of honeycomb-filled round pipes and optimized their crashworthiness design. Inspired by biology, Nian et al. [34] proposed a new type of gradient honeycomb-filled round tube and systematically studied its crashworthiness under lateral load. Yao et al. [35] mainly analyzed the dynamic responses of honeycomb-filled structure under various conditions.

In summary, the novel thin-walled tube obtained by filling with foam and honeycomb material has better crashworthiness. Although there are a large number of studies on the foam or honeycomb filling structures, comparative studies of these two filling methods are rarely reported. Therefore, it is important to conduct in-depth research of the mechanical behavior of the novel thin-walled tubes filled with foam and honeycomb, and then to understand the collision behavior and energy absorption characteristics between them more thoroughly. The difference in the dynamic response of the foam and honeycomb-filled novel thin-walled tubes under different filling styles is systematically studied, specifically involving the energy absorption characteristics, peak impact force, deformation mode and load displacement characteristics. The optimization design of the most promising filling structure with excellent crashworthiness is further conducted to maximize the specific energy absorption and minimize the peak collision force.

2. Numerical Model

2.1. Geometric Model of the Filling Structure

Figure 1 gives the geometric model of the filling structure. R_{inner}, R_{outer} and T are the radius and wall thickness of the inner and outer round tubes, respectively. The dotted line in the picture is the angle bisector of the angle α, and the intersection of two adjacent oblique lines falls on the intersection of the angle bisector and the section line of the outer tube. The sizes of R_{inner}, R_{outer}, T and α are, respectively, 15 mm, 30 mm, 1 mm and 90°. Figure 2b exhibits the different filling methods of these novel thin-walled structures. Among these seven filling styles, G is a full filling style and A–F are partial filling styles. The filling structure adopts the following naming rules: F and H indicate foam and honeycomb, and the second letter indicates the filling style. For example: FA means using foam to fill in A style of filling, HA uses honeycomb to fill in style A.

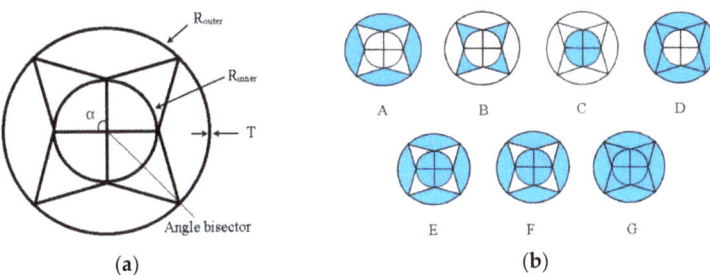

Figure 1. Geometric configuration of the filling structure: (a) novel thin-walled structure; (b) filling styles.

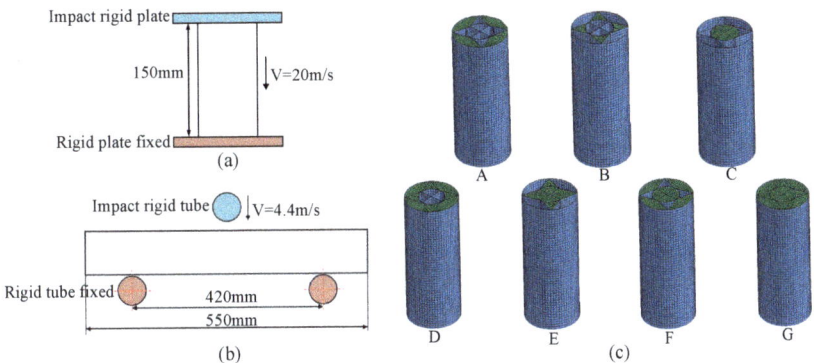

Figure 2. Finite element model of the filled structure. (a) the calculation model of the axial compression; (b) the calculation model of the axial compression; (c) Finite element mesh model of thin-walled tube.

2.2. Finite Element Model

To study the impact behavior of the infill structure under impact load, a model of the infill structure was constructed by the nonlinear dynamic finite element method for simulation. Figure 2 is the numerical simulation model of infill structure. Part a in Figure 2 is the calculation model of the axial compression of the filling structure. The tube wall and honeycomb are treated as shell, the foam is treated as solid. The impact block and rigid wall are set as rigid bodies. When the impact block impacts the thin-walled tube axially at a speed of 20 m/s, the rigid wall at the bottom is fixed. The contact settings in the dynamic compression process are as follows: the thin-walled tube adopts automatic single-surface contact and the contact between thin-walled tube and the rigid wall is set as surface-to-surface contact. The sensitivity analysis of the mesh shows that the mesh size of the element of 1.5 mm × 1.5 mm is sufficient to produce reliable results. In these contacts, the dynamic and static friction coefficients are set to 0.2 and 0.3. The right side of part a in Figure 2 is the model of the filling structure under different filling methods, with a length of 150 mm.

Part b is the finite element model of the filling structure under lateral load. The indenter and the supports are treated as rigid bodies. The indenter impacts the thin-walled structure vertically downward at 4.4 m/s and the supports are fixed during the impact. The settings of contact properties, mesh size and thin-walled tube length are the same as part a.

2.3. Material Properties

The material of the new thin-walled tube and honeycomb filler is aluminum alloy AlMgSi0.5F22 with density $\rho = 2.7 \times 10^3$ kg/m^3, elastic modulus $E = 68.566$ GPa, Poisson's ratio $\nu = 0.29$, yield stress $\sigma_y = 231$ Mpa and ultimate stress $\sigma_{ult} = 254$ Mpa. Considering that aluminum is not sensitive to strain rate, material strain-rate effect can be ignored during simulation analysis [36]. The foam filling is made of foamed aluminum. In order to save calculation cost and ensure sufficient calculation accuracy, crushable foam is used for modeling. The platform stress of lightweight porous materials is very important for its energy absorption. The calculation formula of foam aluminum platform stress σ_p [31,37] is as follows:

$$\sigma_p = C_{pow}(\frac{\rho_f}{\rho_0})^n \tag{1}$$

In the formula, ρ_f and ρ_0 are the density of the foam and foam substrate, respectively. The density of aluminum is $\rho_0 = 2.7 \times 10^3$ kg/m^3. C_{pow} and n are constants. According to the test results in literature [38], $C_{pow} = 526$ Mpa and $n = 2.17$. The simplified functional relationship of the foam stress–strain curve is used for simulation, as shown in Table 1 [39]. The Young's modulus of the foam is $E = 64.8$ GPa, the tensile stress cut-off value is 1.11, the rate-sensitive damping is 0.05 and the Poisson's ratio is 0.01 [31].

Table 1. Simplified stress–strain relationship of aluminum foam.

Strain	0	σ_p/E	0.6	0.7	0.75	0.8
Stress	0	σ_p	σ_p	1.35 σ_p	5 σ_p	0.05 E

2.4. Evaluation Index

Generally, energy absorption (EA), average crushing force (MCF), maximum collision force (MIF), specific energy absorption (SEA) and crushing force efficiency (CLE) are commonly used evaluation indicators. As a key indicator, specific energy absorption (SEA) is often used to evaluate the mechanical performance of thin-walled structures. A larger SEA means better crashworthiness. The calculation formula is as follows:

$$SEA = \frac{EA}{M} \tag{2}$$

Among them, M is the total mass and EA indicates the total absorbed energy of the structure. The calculation formula of EA is as follows:

$$EA = \int_0^S F(x)dx \tag{3}$$

In the formula, S indicates the displacement of impact force and $F(x)$ represents the instantaneous collision force.

CLE is an index for evaluating the uniformity and consistency of the collision force. It is another very important evaluation index for crashworthiness. It can be calculated as:

$$CLE = \frac{MCF}{PCF} \times 100\% \tag{4}$$

Among them, MCF is the average crushing force, PCF is the maximum collision load and the calculation formula of MCF is as follows:

$$MCF = \frac{1}{S}\int_0^S F(x)dx \tag{5}$$

2.5. Validation of the FE Model

To verify the effectiveness of the numerical model, the results of the axial compression simulation of foam-filled multi-cell square tube (F01, F40) are compared with the reference results in literature [40]. Figure 3 shows that the simulation values in this paper are consistent with the results in reference [40], which have been verified by the theoretical results. Subsequently, the bending behavior of foam-filled square tube in reference [41] was simulated. Figure 4 shows the comparison between the test results and the numerical results. Both the force–displacement curve and the deformation mode show a high consistency feature. In summary, the finite element model of axial and lateral impact is sufficiently reliable.

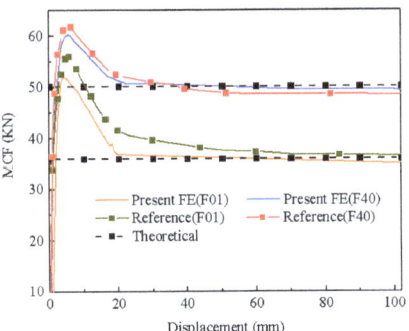

Figure 3. The present FE and reference.

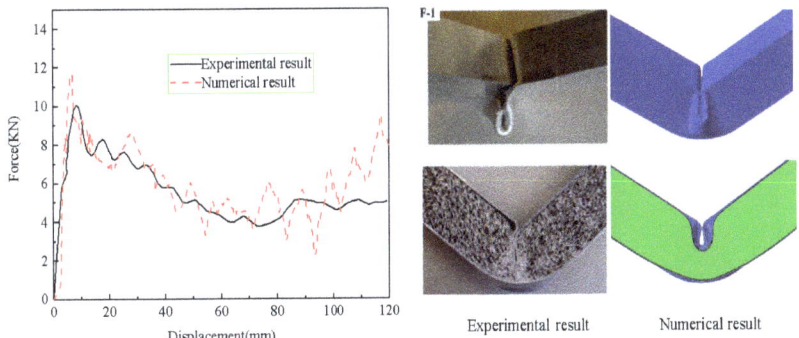

Figure 4. Comparison of test results and numerical results.

3. Numerical Results

3.1. Axial Compression Analysis

In this section, the mechanical behavior of filled structures subjected to axial impact will be studied. The unfilled novel thin-walled tube is also introduced for comparison. Figure 5 shows the EA and SEA of the filling structure in different filling styles. The EA values of the honeycomb- and foam-filled tubes under different filling styles are higher than that of empty tubes (red dotted line) in Figure 5a. In addition, the EA of honeycomb/foam-filled tubes shows obvious differences between different filling styles. Partially filled HB/FB has the smallest EA, which is, respectively, 8.8% and 17.6% higher than that of empty tube. This shows that whether it is honeycomb or foam filling, the different filling methods of novel thin-walled tube are all conducive to total energy absorption. Additionally, the foam-filling method has better enhancement effect.

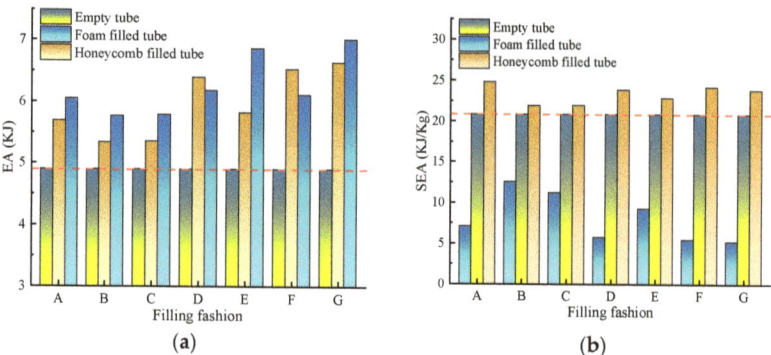

Figure 5. EA and SEA of the filling structure under different filling styles. (a) EA of filling structure. (b) SEA of filling structure.

Considering that foam and honeycomb filling may compromise the weight efficiency of the novel structures, Figure 5b compares the SEA of the filling structures. As shown in the picture, the SEA of the foam-filled tube is relatively low, regardless of the filling method. By contrast, the honeycomb filling method increases the SEA. Among the foam/honeycomb-filled tubes, partially filled FB/HA has the largest SEA, which is 40.4% lower and 18.2% higher than that of the empty 20.86 KJ tube. Although foam filling plays a positive role in improving mechanical properties, it greatly impairs the weight efficiency. It is worth noting that honeycomb filling just compensates for this defect. Figure 6 shows the peak impact force of the filling structure with different filling methods. The foam-filled structure has the greatest peak force, while the empty tube has the least. This shows that honeycomb is more conducive to reducing peak force than foam filling. Among foam/honeycomb-filled tubes, fully-filled FG/HG has the largest PCF, followed by partially filled FE/HD, and partially filled FB/HA is the smallest.

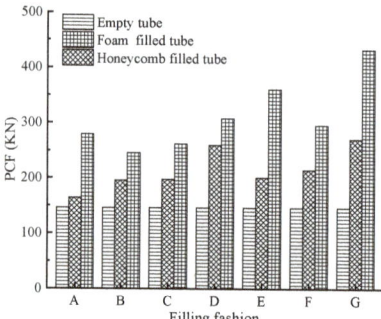

Figure 6. PCF of filling structure under different filling styles.

From the above analysis, we can see that FB/HA has the best crashworthiness among these novel thin-walled tubes. To have a better understand, Figure 7 exhibits deformation modes of FB, HA and the unfilled thin-walled structure. These three thin-walled structures have all undergone orderly and progressive folding. However, they have different folding characteristics. As shown in the partial enlarged cross-sectional view on the right, the honeycomb-filled tube outside (black solid line ellipse) and the number of internal folds (black solid line box) are larger than foam-filled and empty tubes. Although the number of folds on the outside of the foam-filled tube is the same as that of empty tube, there are more folds on the inside of the foam-filled tube. Meanwhile, the whole structure of the foam-filled structure has a certain degree of plastic deformation (such as the blue solid line box) when the compression displacement is 60 mm, while more plastic deformation means that more impact energy is absorbed, which explains why the *EA* of FB is greater than that of HA.

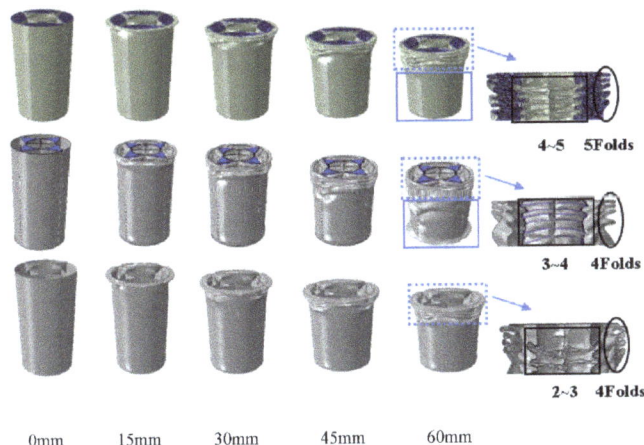

Figure 7. Deformation mode of filling structure.

3.2. Three-Point Bending Analysis

The thin-walled structures will also be subjected to lateral loads in actual use. Thin-walled structures are not only impacted by axial loads, but are sometimes also impacted by lateral loads. Therefore, the bending performance of the structure is very important for its application. Figure 8 shows the *EA*, *SEA*, *PCF* and *CLE* of the filling structure under different filling styles. As shown in Figure 8a, the *EA* of foam- and honeycomb-filled structures are both larger than that of empty tube. Fully filled FG and HG have the largest *EA*, followed by partially filled FA and HD. In particular, under the same filling style, just the *EA* of the honeycomb-filled tube in the filling style C exceeds that of foam-filled tube. This shows that both foam and honeycomb filling will cause the increase in total absorbed energy, and foam filling is more conducive to the growth of *EA* than honeycomb filling. However, it does not represent an increase in its energy absorption efficiency. The *SEA* of honeycomb-filled tube is higher than that of empty tube. The *SEA* of foam/honeycomb-filled pipes showed significant differences. The partially filled FB/HF has the largest *SEA*, which is 27.4%/26.8% lower/higher than the empty tube. The results indicate that the honeycomb filling is an extremely effective means to enhance energy-absorption capacity, and the F filling method may be the best choice.

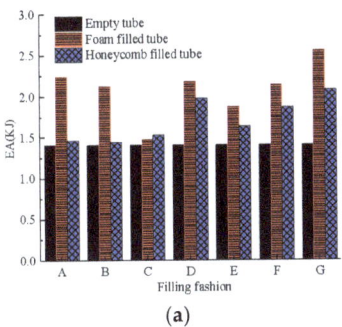
(a)

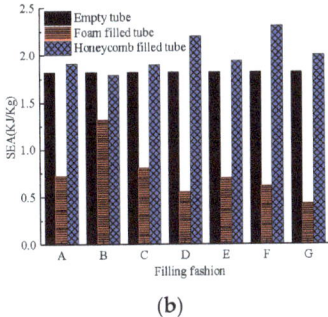
(b)

Figure 8. *Cont.*

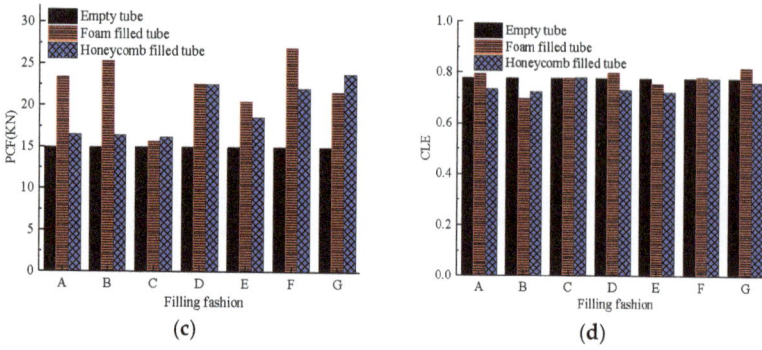

Figure 8. The performance indicatorsof the filling structures under different filling methods: (**a**) *EA*; (**b**) *SEA*; (**c**) *PCF*; (**d**) *CLE*.

The *PCF* of the filling structures is also compared in Figure 8c. The *PCF* of foam and honeycomb filling is greater than that of empty tube, and *PCF* of the foam filling tube is the highest. At the same time, the *PCF* of foam and honeycomb tubes are affected by filling method. In Figure 8d, filling techniques do not make *CLE* of the empty tube change too much and the *CLE* did not change significantly with the change of the filling method.

Based on the above, FB/HF has the best energy absorption characteristics among foam/honeycomb-filled tubes. For further understanding the difference in bending performance, the force–displacement curves, specific energy absorption–displacement curves and deformation modes are selected for comparative analysis, as shown in Figure 9. The impact force for the honeycomb filling tube and the empty tube presents the same trend. It first increases sharply and then slowly decreases, while the collision force of the foam-filled tube shows a monotonous increasing trend. The collision force of foam- and honeycomb-filled structures is bigger, and when the loading displacement is equal to 120 mm, the collision force of foam-filled pipes is the largest. This means that foam and honeycomb-filled tubes can withstand a higher level of lateral impact and transmit greater bending moments. In order to better illustrate this point, Figure 9c shows their deformation modes. It can be seen from the map that the partially recessed area of the foam-filled tube is arc-shaped. The effective contact area is larger than that of the honeycomb-filled and empty tube. Although honeycomb filling and empty tube both have the phenomenon of concentrated deformation area, the local recessed area is V-shaped. However, the cross-sectional deformation of the partially recessed area of the honeycomb-filled tube is more obvious than that of empty tube. As shown in the *SEA*-displacement curve in Figure 9b, the *SEA* of honeycomb-filled tube is the largest. This is just the opposite for foam-filled tube. Therefore, foam-filled tubes are not the best choice for crashworthiness.

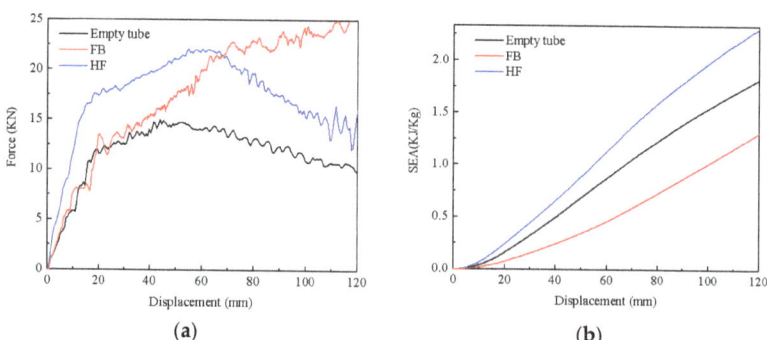

Figure 9. *Cont.*

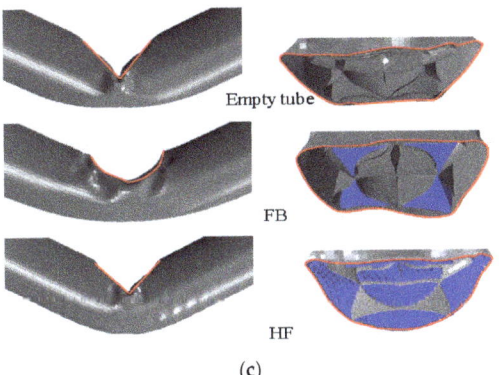

Figure 9. Comparison of filled structures: (a) force–displacement diagram; (b) SEA-displacement diagram; (c) deformation mode.

4. Multi-Objective Optimization Design

4.1. Optimization Problem Set-Up

It is often required that thin-walled structures can absorb the most energy in a certain range of peak stress. Therefore, SEA and PCF are selected as two objectives of this optimal design. Crashworthiness optimization aims to maximize SEA and minimize PCF. However, SEA and PCF are in conflict with each other. Therefore, a multi-objective optimization design is selected to solve this contradictory objective problem [42–45]. From the analysis in Section 3.2, FB and HF have better crash resistance. Moreover, the crashworthiness of HF is better than that of FB. However, the optimal crashworthiness of these structures still depends on different structure and material parameters. Therefore, in this section, the novel thin-walled tube wall thickness T, honeycomb wall thickness t and foam density ρ_f are used as design variables. The crashworthiness optimization problem is described as follows:

The optimized expression of FB is as follows:

$$\begin{cases} \text{Minimize } [PCF(T, \rho_f), -SEA(T, \rho_f)] \\ s.t \begin{cases} 0.5 \text{ mm} \leq T \leq 1.5 \text{ mm} \\ 170 \text{ Kg/m}^3 \leq \rho_f \leq 340 \text{ Kg/m}^3 \end{cases} \end{cases} \quad (6)$$

The optimized expression of HF is as follows:

$$\begin{cases} \text{Minimize } [PCF(T, \rho_f), -SEA(T, \rho_f)] \\ s.t \begin{cases} 0.5 \text{ mm} \leq T \leq 1.5 \text{ mm} \\ 0.01 \text{ mm} \leq t \leq 0.1 \text{ mm} \end{cases} \end{cases} \quad (7)$$

The optimized expression of empty tube is as follows:

$$\begin{cases} \text{Minimize } [PCF(T, \rho_f), -SEA(T, \rho_f)] \\ s.t \{ 0.5 \text{ mm} \leq T \leq 1.5 \text{ mm} \end{cases} \quad (8)$$

4.2. Experimental Design

The main experimental design methods include central composite design, Taguchi orthogonal experiment method [45], Latin hypercube design and full-factor experiment. Because the full-factor experiment has good uniformity [10,39], this paper uses the full-factor design to generate 16 sample points (four levels for design variables T, t, ρ_f), as shown in Figure 10. Subsequently, numerical simulation on these sample points is carried out and corresponding response values are obtained.

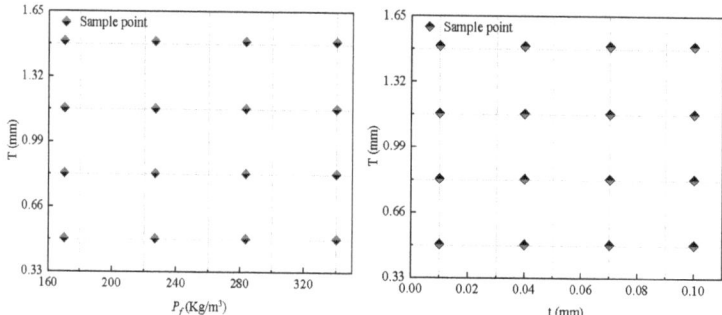

Figure 10. Design sample points of FB and HF.

4.3. Predictive Model

Since the construction of SEA and PCF forecasting models is important for the crashworthiness optimization, the accuracy of forecasting models needs to be verified. Firstly, through polynomial regression analysis (PR), the functional relationship between the optimization objective and design parameters is established, and the corresponding prediction model is obtained. Then, the square value of index R (R^2), adjusted R^2 (R^2_{adj}), root mean square error (RMSE) and maximum relative error (MARE) are used to evaluate the accuracy of this prediction model. The corresponding expression is as follows:

$$R^2 = 1 - \frac{\sum_{i=1}^{n}(y_i - \hat{y}_i)^2}{\sum_{i=1}^{n}(y_i - \bar{y})^2} \tag{9}$$

$$R^2_{adj} = 1 - (1 - R^2)\frac{n-1}{n-k-1} \tag{10}$$

$$RMSE = \sqrt{\frac{\sum_{i=1}^{n}(y_i - \hat{y}_i)^2}{n}} \tag{11}$$

$$MARE = \max_{i=1,2,\ldots n}(\frac{|y_i - \hat{y}_i|}{|y_i|}) \tag{12}$$

Among them, y_i represents the value of the design points obtained by experiment and numerical analysis, $\hat{y}_i$ is the predicted value of prediction model, n is the number of experimental sample points, $\bar{y}$ represents the average value of y_i and k is the number of non-constant items. Normally, the closer R^2 is to 1, the higher the degree of fit; the smaller the RSME and MARE, the more accurate the prediction model. Table 2 gives the accuracy index of the forecasting model of FB, HF and empty tube. From Table 2, we can see that all R^2 values are close to 1, and all MARE values are less than 6%. Therefore, it can be considered that these PR mathematical models are accurate enough to be used in crashworthiness optimization.

Table 2. Accuracy of the prediction model.

Objectives	SEA				PCF			
Estimators	R^2	R^2_{adj}	MARE	RMSE	R^2	R^2_{adj}	MARE	RMSE
FB	0.9891	0.98	2.39%	0.026	0.9949	0.9907	2.18%	1.291
HF	0.9784	0.9604	2.61%	0.0307	0.9558	0.9352	5.08%	0.3869
Empty tube	0.9987	0.9977	0.89%	0.0115	0.9981	0.9965	1.32%	1.1887

4.4. Particle Swarm Algorithm and Optimization Process

Since the particle swarm algorithm has the advantages of easy implementation, high accuracy and fast convergence [46,47], this paper uses the particle swarm algorithm to obtain a Pareto relatively optimal solution of the prediction model. Ten particles are set for tracking and each particle moves 100 times in the constrained space for accuracy. The inertia weight w is an important parameter that affects the pros and cons of the particle swarm algorithm. The solution of the HF prediction model obtained by the particle swarm algorithm under different inertia weights is shown in Figure 11. It can be seen from the picture that when the inertia weight is equal to 0.7, the solutions of the *SEA* and *PCF* prediction models have undergone large oscillations at first, and then stabilized in a certain area. Therefore, when the inertia weight is equal to 0.7, the convergence is best. The specific parameters of the particle swarm algorithm are shown in Table 3. Figure 12 is the flow chart of the optimized design. First, the full-factor experimental design is carried out on the problem of clear optimized design. Further, perform simulation analysis on the sample points to obtain the target response value. Then, the prediction models of *SEA* and *PCF* were constructed through polynomial regression analysis. Finally, the Pareto optimal solution set is given after the calculation of the prediction model by the optimized algorithm.

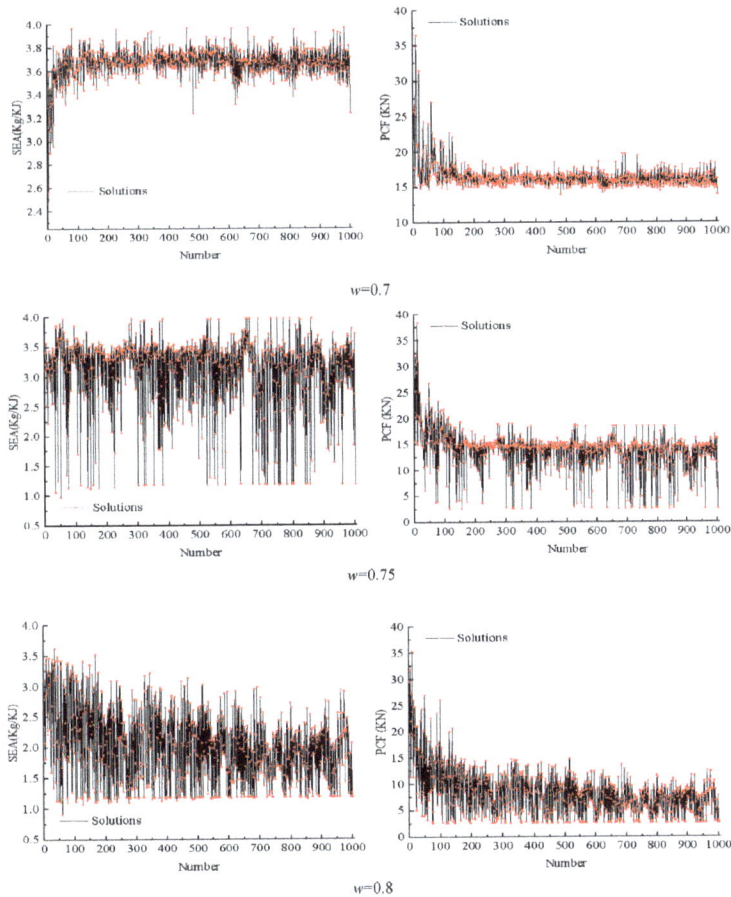

Figure 11. The solutions of *SEA* and *PCF* under different inertia weights.

Table 3. Parameters of particle swarm algorithm.

Parameters	Value
Number of particles	10
Maximum number of iterations	100
Inertia weight	0.7
Personal learning coefficient	1.5
Global learning coefficient	1.5

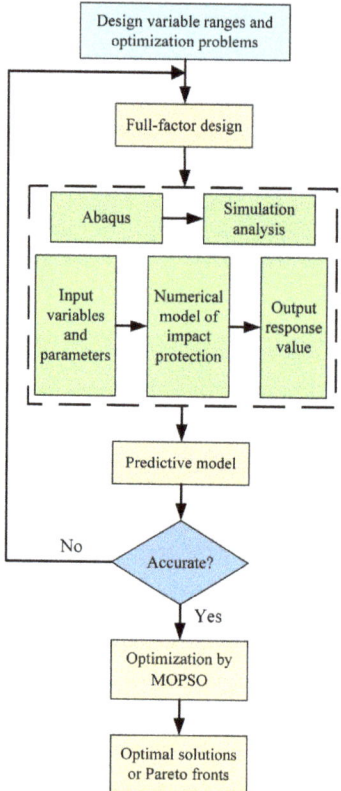

Figure 12. Optimization design flow chart.

4.5. Multi-Objective Optimization Results

To comprehensively analyze the crashworthiness difference between FB, HF and empty tube more, the Pareto boundary obtained after particle swarm optimization is compared and shown in Figure 13. It can be found from the figure that when the PCF is constant, the closer the Pareto boundary is to the left, the greater the specific energy absorption (SEA). The Pareto optimal solution set of HF is closest to the left, followed by the empty tube. Therefore, the crashworthiness of HF is better than that of FB and empty tube, and the crashworthiness of FB is the worst. In engineering applications, designers can choose based on requirements of PCF. When the PCF is less than or equal to 15 KN, the red five-pointed star in the map is the optimal design point of each structure. Figure 14 shows the force–displacement diagrams and deformation modes of these three optimal structures. The PCF of the three structures in Figure 14a is less than 15 KN, and HF has the largest PCF. In Figure 14b, local recessed area of the HF optimal structure presents an arc shape, while both FB and empty tube present V shapes.

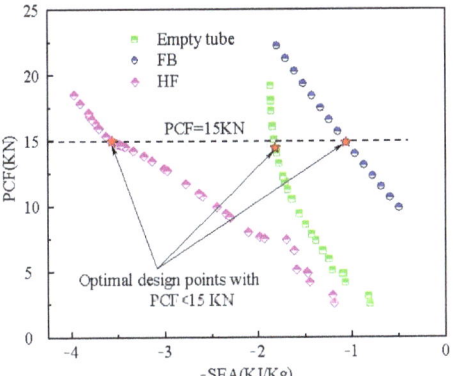

Figure 13. Comparison of the Pareto boundary of FB, HF and empty tube.

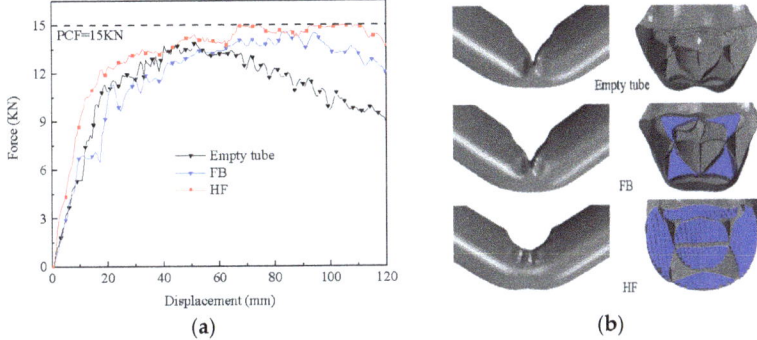

Figure 14. Comparison of the optimal structure: (**a**) force–displacement diagram; (**b**) deformation mode.

5. Conclusions

To further enhance the crashworthiness of the novel thin-walled structure, foam and honeycomb are used to fill it. The numerical simulation model of the filled structure was first built and verified using the nonlinear dynamic Abaqus software. Then, the impact resistance of honeycomb-filled tubes, foam-filled tubes and empty tube under axial load was systematically compared and analyzed. Furthermore, based on the force–displacement curve, specific energy absorption and deformation model, a comparative analysis of the mechanical behavior of filled tubes subjected to lateral impact was carried out. The optimization design of the most promising filling structure with excellent crashworthiness was finally conducted to maximize the specific energy absorption and minimize the peak collision force. The results of this study show that:

(1) The introduction of honeycomb filling and foam filling enabled the thin-walled structures to absorb more energy. The total absorbed energy increases at least 8.8% compared with the empty tube. At the same time, the crashworthiness of the filling structure was closely related to the filling styles. The foam filling will greatly impair the weight efficiency of the novel thin-walled tube. However, honeycomb filling was beneficial to the improvement of *SEA*, which can be improved by up to 18.2%.

(2) Honeycomb filling was more conducive to the reduction of *PCF* than foam filling under axial load. Among foam/honeycomb-filled tubes, fully filled FG/HG had the largest *PCF*, followed by partially filled FE/HD and FB/HA was the smallest.

(3) Under the action of lateral load, foam- and honeycomb-filled tubes could withstand a higher level of lateral impact and transmit greater bending moments than empty tube.

Among foam/honeycomb-filled tubes, FB/HF was the most promising structure with excellent crashworthiness.

(4) The particle swarm algorithm was further used for crashworthiness optimization design of FB and HF, and the Pareto boundaries were obtained and compared. By way of contrast, the optimal structure of HF showed the best crashworthiness. In practical engineering applications, the use of honeycomb-filled novel thin-walled tube may be the best choice.

Author Contributions: Conceptualization, Y.T.; resources, L.L.; software, Y.W.; writing—review and editing, Y.W. and D.X.; methodology, Q.H. All authors have read and agreed to the published version of the manuscript.

Funding: This research is supported by The National Natural Science Foundation of China (No. 51705215), The Chinese Postdoctoral Science Foundation (2022M712932) and the Natural Science Fundamental Research Project of Jiangsu Universities (No. 22KJA460003).

Institutional Review Board Statement: Not applicable.

Informed Consent Statement: Not applicable.

Data Availability Statement: Not applicable.

Conflicts of Interest: The authors declare there is no conflict of interest regarding the publication of this work.

References

1. Abramowicz, W.; Jones, N. Dynamic axial crushing of circular tubes. *Int. J. Impact Eng.* **1984**, *2*, 263–281. [CrossRef]
2. Kim, H.S. New extruded multi-cell aluminum profile for maximum crash energy absorption and weight efficiency. *Thin-Walled Struct.* **2002**, *40*, 311–327. [CrossRef]
3. Najafi, A.; Rais-Rohani, M. Mechanics of axial plastic collapse in multi-cell, multi-corner crush tubes. *Thin-Walled Struct.* **2011**, *49*, 1–12. [CrossRef]
4. He, Q.; Wang, Y.; Gu, H.; Feng, J.; Zhou, H. The dynamic behavior of fractal-like tubes with Sierpinski hierarchy under axial loading. *Eng. Comput.* **2022**, *38*, 1285–1298. [CrossRef]
5. Zhang, X.; Zhang, H. Axial crushing of circular multi-cell columns. *Int. J. Impact Eng.* **2014**, *65*, 110–125. [CrossRef]
6. Galib, D.A.; Limam, A. Experimental and numerical investigation of static and dynamic axial crushing of circular aluminum tubes. *Thin-Walled Struct.* **2004**, *42*, 1103–1137. [CrossRef]
7. Zhang, X.; Cheng, G.; Zhang, H. Theoretical prediction and numerical simulation of multi-cell square thin-walled structures. *Thin-Walled Struct.* **2006**, *44*, 1185–1191. [CrossRef]
8. Zhang, X.; Zhang, H. Energy absorption of multi-cell stub columns under axial compression. *Thin-Walled Struct.* **2013**, *68*, 156–163. [CrossRef]
9. Nia, A.A.; Parsapour, M. Comparative analysis of energy absorption capacity of simple and multi-cell thin-walled tubes with triangular, square, hexagonal and octagonal sections. *Thin-Walled Struct.* **2014**, *74*, 155–165.
10. Yin, H.; Xiao, Y.; Wen, G.; Qing, Q.; Wu, X. Crushing analysis and multi-objective optimization design for bionic thin-walled structure. *Mater. Des.* **2015**, *87*, 825–834. [CrossRef]
11. Zhu, B.; Zhang, M.; Zhao, J. Micro structure and mechanical properties of sheep horn. *Microsc. Res. Tech.* **2016**, *79*, 664–674. [CrossRef] [PubMed]
12. Song, J.F.; Xu, S.C.; Wang, H.X.; Wu, X.Q.; Zou, M. Bionic design and multi-objective optimization for variable wall thickness tube inspired bamboo structures. *Thin-Walled Struct.* **2018**, *125*, 76–88. [CrossRef]
13. Jiang, H.; Ren, Y.; Jin, Q.; Hu, Y.; Cheng, F. Crashworthiness of novel concentric auxetic reentrant honeycomb with negative Poisson's ratio biologically inspired by coconut palm. *Thin-Walled Struct.* **2020**, *154*, 106911. [CrossRef]
14. Song, J.; Xu, S.; Liu, S.; Huang, H.; Zou, M. Design and numerical study on bionic columns with grooves under lateral impact. *Thin-Walled Struct.* **2020**, *148*, 106546. [CrossRef]
15. Zou, M.; Xu, S.; Wei, C.; Wang, H.; Liu, Z. A bionic method for the crashworthiness design of thin-walled structures inspired by bamboo. *Thin-Walled Struct.* **2016**, *101*, 222–230. [CrossRef]
16. Palombini, F.L.; Mariathb, J.E.A.; Oliveiraa, B.F. Bionic design of thin-walled structure based on the geometry of the vascular bundles of bamboo. *Thin-Walled Struct.* **2020**, *155*, 106936. [CrossRef]
17. Ferdynus, M.; Rozylo, P.; Rogala, M. Energy Absorption Capability of Thin-Walled Prismatic Aluminum Tubes with Spherical Indentations. *Materials* **2020**, *13*, 4304. [CrossRef]
18. Orbulov, I.N.; Májlinger, K. Characteristic compressive properties of hybrid metal matrix syntactic foams. *Mater. Sci. Eng. A* **2014**, *606*, 248–256.

19. Orbulov, I.N.; Szlancsik, A.; Kemény, A.; Kincses, D. Compressive mechanical properties of low-cost, aluminium matrix syntactic foams. *Compos. A Appl. Sci. Manuf.* **2020**, *135*, 105923. [CrossRef]
20. Szlancsik, A.; Orbulov, I.N. Compressive properties of metal matrix syntactic foams in uni- and triaxial compression. *Mater. Sci. Eng. A* **2021**, *827*, 142081. [CrossRef]
21. Movahedi, N.; Vesenjak, M.; Krstulovi-Opara, L.; Belova, I.V.; Murch, G.E.; Fiedler, T. Dynamic compression of functionally-graded metal syntactic foams. *Compos. Struct.* **2021**, *261*, 113308. [CrossRef]
22. Marx, J.; Rabiei, A. Study on the Microstructure and Compression of Composite Metal Foam Core Sandwich Panels. *Metall. Mater. Trans. A* **2020**, *51*, 5187–5197. [CrossRef]
23. Marx, J.; Rabiei, A. Tensile properties of composite metal foam and composite metal foam core sandwich panels. *J. Sandw. Struct. Mater.* **2021**, *23*, 3773–3793. [CrossRef]
24. Sahil, G.; Anand, C.S.; Sunil, K.S.; Rakesh, C.S. Crashworthiness analysis of foam filled star shape polygon of thin-walled structure. *Thin-Walled Struct.* **2019**, *144*, 106312.
25. Zhang, Y.; Sun, G.; Li, G.; Luo, Z.; Li, Q. Optimization of foam-filled bitubal structures for crashworthiness criteria. *Mater. Des.* **2012**, *38*, 99–109. [CrossRef]
26. Gao, Q.; Wang, L.; Wang, Y.; Wang, C. Crushing analysis and multiobjective crashworthiness optimization of foam-filled ellipse tubes under oblique impact loading. *Thin-Walled Struct.* **2016**, *100*, 105–112. [CrossRef]
27. Hanssen, A.; Langseth, M.; Hopperstad, O. Optimum design for energy absorption of square aluminium columns with aluminium foam filler. *Int. J. Mech. Sci.* **2001**, *43*, 153–176. [CrossRef]
28. Yin, H.; Wen, G.; Wu, X.; Qing, Q.; Hou, S. Crashworthiness design of functionally graded foam-filled multi-cell thin-walled structures. *Thin-Walled Struct.* **2014**, *85*, 142–155. [CrossRef]
29. Li, Z.B.; Zheng, Z.J.; Yu, J.L.; Guo, L.W. Crashworthiness of foam-filled thin-walled circular tubes under dynamic bending. *Mater. Des.* **2013**, *52*, 1058–1064. [CrossRef]
30. Li, Z.B.; Lu, F.Y. Bending resistance and energy-absorbing effectiveness of empty and foam-filled thin-walled tubes. *J. Reinf. Plast. Compos.* **2015**, *34*, 761–768. [CrossRef]
31. Qi, C.; Sun, Y.; Yang, S. A comparative study on empty and foam-filled hybrid material double-hat beams under lateral impact. *Thin-Walled Struct.* **2018**, *129*, 327–341. [CrossRef]
32. Pandarkar, A.; Goel, M.D.; Hora, M.S. Axial crushing of hollow and foam filled tubes: An overview. *Sadhana Acad. Proc. Eng. Sci.* **2016**, *41*, 909–921. [CrossRef]
33. Cakıroglu, C. Quasi-Static Crushing Behavior of Nomex Honeycomb Filled Thin-Walled Aluminum Tubes. Ph.D. Thesis, Izmir Institute of Technology, Izmir, Turkey, 2011; pp. 184–193.
34. Nian, Y.Z.; Wan, S.; Lia, X.Y. How does bio-inspired graded honeycomb filler affect energy absorption characteristics. *Thin-Walled Struct.* **2019**, *144*, 106269. [CrossRef]
35. Yao, S.; Xiao, X.; Xu, P.; Qu, Q.; Che, Q. The impact performance of honeycomb-filled structures under eccentric loading for subway vehicles. *Thin-Walled Struct.* **2018**, *123*, 360–370. [CrossRef]
36. Chen, W. Experimental and numerical study on bending collapse of aluminum foam-filled hat profiles. *Int. J. Solids Struct.* **2001**, *38*, 7919–7944. [CrossRef]
37. Yang, S.; Qi, C. Multiobjective optimization for empty and foam-filled square columns under oblique impact loading. *Int. J. Impact Eng.* **2013**, *54*, 177–191. [CrossRef]
38. Hanssen, A.G.; Hopperstad, O.S.; Langseth, M.; Ilstad, H. Validation of constitutive models applicable to aluminium foams. *Int. J. Mech. Sci.* **2002**, *44*, 359–406. [CrossRef]
39. Hou, S.; Li, Q.; Long, S.; Yang, X.; Li, W. Crashworthiness design for foam filled thin-wall structures. *Mater. Des.* **2009**, *30*, 2024–2032. [CrossRef]
40. Zhang, Y.; Ge, P.Z.; Lu, M.H.; Lai, X.G. Crashworthiness study for multi-cell composite filling structures. *Int. J. Crashworthiness* **2018**, *23*, 32–46. [CrossRef]
41. Zarei, H.R.; Kroger, M. Bending behavior of empty and foam-filled beams: Structural optimization. *Int. J. Impact Eng.* **2008**, *35*, 521–529. [CrossRef]
42. Pirmohammad, S.; Esmaeili-Marzdashti, S. Multi-objective crashworthiness optimization of square and octagonal bitubal structures including different hole shapes. *Thin-Walled Struct.* **2019**, *139*, 126–138. [CrossRef]
43. Ying, L.; Dai, M.H.; Zhang, S.Z.; Ma, H.L.; Hu, P. Multiobjective crashworthiness optimization of thin-walled structures with functionally graded strength under oblique impact loading. *Thin-Walled Struct.* **2017**, *117*, 165–177. [CrossRef]
44. Deng, X.L.; Liu, W.Y. Multi-objective optimization of thin-walled sandwich tubes with lateral corrugated tubes in the middle for energy absorption. *Thin-Walled Struct.* **2019**, *137*, 303–317. [CrossRef]
45. Fang, H.; Rais-Rohani, M.; Liu, Z.; Horstemeyer, M.F. A comparative study of metamodeling methods for multiobjective crashworthiness optimization. *Comput. Struct.* **2005**, *83*, 2121–2136. [CrossRef]
46. Wang, T.; Li, Z.; Wang, L.; Hulbert, G.M. Crashworthiness analysis and collaborative optimization design for a novel crash-box with re-entrant auxetic core. *Struct. Multidiscip. Optim.* **2020**, *62*, 2167–2179. [CrossRef]
47. Su, L.W.; Zhang, Y.J.; Sun, B.B. Multi-objective optimization of deployable composite cylindrical thin-walled hinges with progressive damage. *Struct. Multidiscip. Optim.* **2019**, *61*, 803–817. [CrossRef]

Article

Microstructural Evolution and Mechanical Properties of Al-Si-Mg-Cu Cast Alloys with Different Cu Contents

Pengfei Zhou [1,2,3], Dongtao Wang [1,2,*], Hiromi Nagaumi [1,2,*], Rui Wang [1,2], Xiaozu Zhang [1,2], Xinzhong Li [1,2], Haitao Zhang [1,4] and Bo Zhang [5]

1. School of Iron and Steel, Soochow University, Suzhou 215021, China
2. High-Performance Metal Structural Materials Research Institute, Soochow University, Suzhou 215021, China
3. School of Automotive and Transportation, Yancheng Polytechnic College, Yancheng 215400, China
4. Key Laboratory of Electromagnetic Processing of Materials, Ministry of Education, Northeastern University, Shenyang 110819, China
5. Shandong Weiqiao Aluminum & Electricity Co., Ltd., Binzhou 256200, China
* Correspondence: dtwang@suda.edu.cn (D.W.); zhanghai888jp@suda.edu.cn (H.N.)

Abstract: The mechanical properties of Al-Si-Mg-Cu cast alloys are heavily determined by Cu content due to the precipitation of relating strengthening precipitates during the aging treatment. In this study, the microstructures and mechanical properties of Al-9Si-0.5Mg-xCu (x = 0, 0.9, 1.5, and 2.1 wt.%) alloys were investigated to elucidate the effect of Cu content on the evolution of their mechanical properties. After T6 (480 °C + 6 h − 530 °C + 4 h, 175 °C + 10 h) treatment, Mg-rich and Cu-rich phases were dissolved in the matrix; the main aging-precipitates of the alloys change from the needle-like β″ phases in the base alloy to the granular Q′ phases in the 0.9Cu alloy, the granular Q′ phase in the 1.5Cu alloy, the granular Q′ phase, and θ′ platelets in the 2.1Cu alloy. The increase of Cu level results in difference of the type, number density, and morphology of the nanoscale precipitated phase. Because of precipitation strength, the yield strength was increased by 103–130 MPa depending on the Cu contents. The precipitation strengthening effect of the precipitates was quantitatively evaluated by the Orowan mechanism. The aging-treated Al-9Si-0.5Mg-2Cu alloy shows the good strength and ductility: yield strength 351 MPa, ultimate tensile strength 442 MPa, and elongation 8.4%. The morphologies of fracture surfaces of the alloys also were observed.

Keywords: Al-Si-Mg-Cu alloy; Cu content; microstructure; precipitate; mechanical properties

1. Introduction

Al-Si cast aluminum alloys are extensively used in the automobile field due to their superior castability, satisfactory mechanical and physical properties, and low coefficient of thermal expansion [1–4]. Adding Mg to Al-Si alloys plays a role in solid solution strengthening and precipitation hardening of aging treatment [5–7]. Cu can significantly increase the mechanical properties of Al-Si-Mg alloys with the formation of nanoscale θ′ and Q′ precipitates during aging [7–10]. Unlike that of Al-Si-Mg and Al-Si-Cu cast alloys, the precipitation strengthening of the Al-Si-Mg-Cu alloy is mainly related to the β″, θ′, and/or Q′ phases; meanwhile, the type and volume fraction of Mg and/or Cu-rich precipitates are closely related to heat-treatment conditions and Cu level [7,9,10]. However, the enhanced strength of the Al-Si-Mg alloys with Cu addition is usually at the expense of their ductility. In addition, the addition of Cu can decrease the melting point and eutectic temperature of Al-Si-Mg alloys, leading to an increase in the solidification range of the alloys and facilitating porosity formation. Simultaneously, with the increase in Cu content, the precipitates in Al-Si-Mg alloys constantly change in type, morphology, quantity, and size [7,9–13]. Some useful understandings have been reported in the properties and precipitation behavior change with Cu addition in these quaternary alloys [4,7,8]. Shang [12] systematically analyzed the phase component of these alloys with a wide Cu level (0.01–4.5 wt.%) and discussed the effect of precipitates on mechanical properties. The previous works mainly focus on the strengthen effect of θ' precipitates in high-Cu/low-Mg Al-Si-Mg-Cu alloy. When Mg content increases and ratio of Cu/Mg decreases, the Q' nano-phase may show a high fraction after aging treatment, which may change the aging-strengthen effect. However, the work on systematic observation and characterization needs to be conducted in greater detail in low-ratio of Cu/Mg alloy, including the effects of precipitates on the hardening behavior of these alloys.

Therefore, the current study mainly evaluates the effect of Cu content on the microstructural evolution and mechanical properties of Al-9Si-0.5Mg-xCu alloys (Cu/Mg: 0–4) and discusses the contribution of precipitates to the hardening behavior of Al-Si-Mg-xCu alloys.

2. Materials and Methods

2.1. Material Preparation

Commercial-purity Al (99.9%), pure Mg (99.95%), pure Cu (99.9%), Al-20%Si master alloys (all compositions quoted in this article are in weight percentage unless otherwise mentioned), and Al-10%Sr alloy as metamorphic eutectic Si were used to prepare Al-Si-Mg-Cu alloys in a 50 kW resistance furnace. Commercial-purity Al and Al-20%Si were first melted in the resistance furnace. The melt was heated to 740 °C and held at that temperature for 30 min to ensure all components were sufficiently mixed. Then, pure Cu and Mg were added into the melt at 750 °C and held for 20 min, followed by slag removal. The Al-10%Sr alloy was added into alloy melts at 730 °C. The alloy melt was ultimately poured into a water-cooled copper mold (25 × 100 × 200 mm^3, Figure 1) to form an as-cast ingot [14–16].

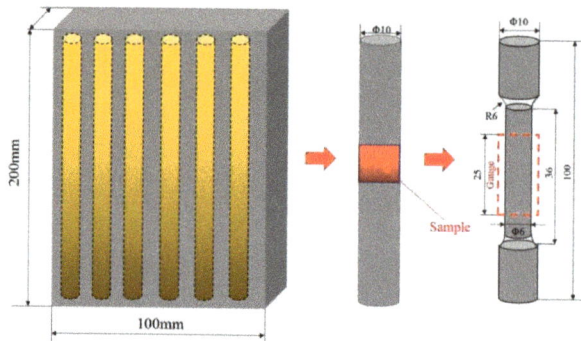

Figure 1. Schematic of the sample position for microstructural and tensile tests.

2.2. Material Characterization

The measured compositions of the designed Al-Si-Mg-xCu alloys, which were noted as A1, A2, A3, and A4 were measured using an SPECTROLAB stationary metal analyzer (SPECTROLAB M12, Kleve, Germany). The results are listed in Table 1.

Table 1. Composition of the Al-Si-Mg-Cu alloys [wt.%].

Alloy	Si	Mg	Cu	Fe	Sr	Other	Al
A1	8.53	0.43	0.01	0.116	0.0195	<0.01	Bal.
A2	8.52	0.42	0.89	0.116	0.0196	<0.01	Bal.
A3	8.51	0.44	1.43	0.117	0.0219	<0.01	Bal.
A4	8.54	0.43	2.08	0.117	0.0211	<0.01	Bal.

Compositional analysis and microstructural evaluation were conducted on samples near the center of the Φ10 mm tensile rods (the red area in Figure 1). The phase compositions of the as-cast alloys were analyzed using a X-ray diffraction (XRD) to identify the phase composition of the alloys with CuKα1 radiation by using PW3040/60X diffractometers. The samples were etched for 2–10 s by using 0.5% hydrofluoric acid for scanning electron microscopy (SEM) characterization using a Phenom X1 scanning electron microscopy (SEM) equipped with X-ray energy dispersive spectroscopy (EDS). The secondary dendrite arm spacing (SDAS) was measured by the intercept method. Quantitative measurements of the SDAS were conducted by optical microscopy using image analysis software (MEDIA CYBERNETICS, Rockville, MD, USA), at least 50 dendrites were measured and their average value is considered as the representation of SDAS [17].

Transmission electron microscope (TEM) samples of the region near the fracture were prepared by sectioning the tensile specimens in the transverse direction. The section near the center of the specimen was polished by hand to approximately 50 μm before a standard 3 mm disc was punched out. Then, the samples were placed on a Gatan 695 PIPS ion beam thinner (Gatan, Inc., Pleasanton, CA, USA). A FEI Tolos F200x (TEM, Tolos F200x, FEI Ltd., Pleasanton, CA, USA) transmission electron microscope equipped with the energy dispersive X-ray spectrometer was operated at an accelerating voltage of 200 kV. All images were taken along the <001>Al zone axis in order to characterize the cross-sections and side views of the precipitates. The average length (l) was calculated using 500 precipitates growing along [100] Al and [010] Al in total. The average area of cross-section of the precipitates (Acs) were calculated in 60 HRTEM images.

2.3. Mechanical Testing

All samples were treated with the solution at 480 °C − 6 h + 530 °C − 4 h, followed by cold water quenching (about 20 °C). Aging treatments were then performed at 175 °C for 10 h. The tensile property of the samples was tested on a DNS-300 universal experimental machine produced by Changchun Machinery Research Institute at a tensile rate of 1 mm/min. The extensometer with a gauge length of 25 mm was used. At least five tensile test specimens were tested for each alloy.

3. Results and Discussions

3.1. As-Cast Microstructures

Figure 2 presents the backscattered SEM images of the as-cast alloys. All alloys have α-Al dendritic microstructure, eutectic Si, and eutectic Mg/Cu-containing phases. Secondary dendrite arm spacing was calculated by Image-Pro Plus (6.0, Media Cybernetics, Inc, Rockville, MD, USA), and the values of the A1–A4 alloys were 22.76 mm, 18.66 mm, 18.02 mm, and 18.25 mm, respectively. As shown in Figure 2a, in the absence of Cu, several black Chinese character-shaped phases are present in the as-cast A1 alloy, and the Energy Dispersive Spectrometer (EDS) result indicates that the composition of the back phase is Al-1.79 at.%Mg-6.06 at.%Si, indicating the Mg_2Si phase [4,5]. With Cu content increasing to 0.9 wt.%, the quantity of Mg_2Si phase decreases. The bright phases are observed in the A2 alloy. The bright phases are dispersed in the eutectic silicon region (Figure 2b). The EDS analysis indicates that the bright phase is Al-15.69 at.%Cu-12.46 at.%Mg-12.05 at.%Si. In A3 alloy, the bright phases increase and the Al_2Cu phases are observed, the composition is Al-29.01 at.%Cu. With the Cu content further increasing from 1.5 to 2.1 wt.%, the bright Al_2Cu phase increases (Figure 2d). Moreover, the Fe-containing phases were observed in the four alloys (Figure 2), the composition of the Fe-rich phase are: Al-26.38 at.%Si-12.13 at.%Mg-3.78 at.%Fe in A1 alloy, Al-26.22 at.%Si-14.76 at.%Mg-4.29 at.%Fe in A2 alloy, Al-25.45 at.%Si-17.19 at.%Mg-5.59 at.%Fe in A3 alloy, and Al-27.56 at.%Si-11.35 at.%Mg-3.42 at.%Fe in A4 alloy.

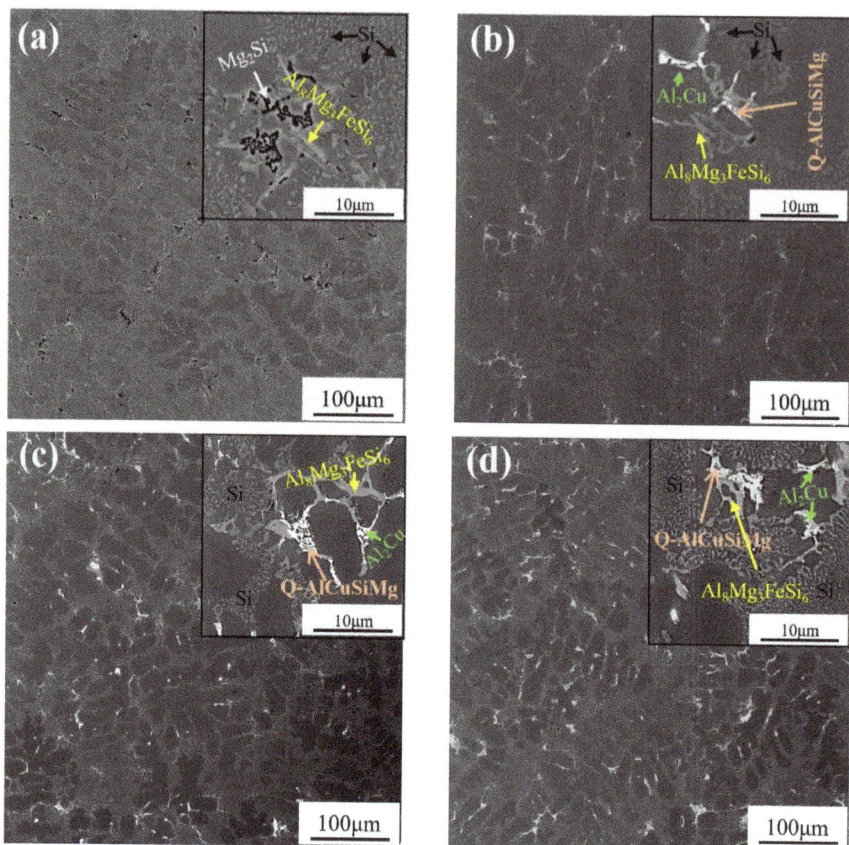

Figure 2. SEM-BSE images of as-cast alloy (**a**) A1, (**b**) A2, (**c**) A3, (**d**) A4.

XRD patterns of the as-cast alloys are presented in Figure 3. The A1 alloy consisted of α-Al, Si, and Mg$_2$Si phases, which is consistent with the microstructural observation in Figure 2a. Compared with that of the A1 alloy, in A3 and A4 alloys, the diffraction peaks of Q-AlCuMgSi and θ-Al$_2$Cu phases were observed, indicating that the Cu addition results in the formation of Q and θ phases. These results are consistent with Figure 2c,d. Therefore, the phase composition of the Al-Si-Mg-Cu alloy system was closely related to the Cu content; meanwhile, the content of each phase was also determined by the Cu and Mg contents [6–10]. In A2 alloys, the XRD results do not show the diffraction peaks of Q and θ phases, but the microstructure in Figure 2b indicates the presence of Q and θ phases. The low fractions of Q and θ phases in low-Cu A2 alloy may result that they hardly be detected by XRD. Moreover, the XRD results indicate the presence of Al$_8$Mg$_3$FeSi$_6$ phase in A1–A4 alloys.

3.2. Microstructures after T6 Treatment

Heat treatment can affect the microstructural features of Al-Si-Cu-Mg alloys [4,11]. Figure 4 presents the SEM-mapping images of the alloys after T6 treatment. Compared with the as-cast state (Figure 2), the Cu-containing phases were mostly dissolved into the α-Al matrix after solid solution treatment. The residual bright phases after solution treatment are mainly Fe-containing phases, which show the same distribution of Mg and Fe in Figure 4a,b and can be identified as the AlFeMgSi phase. In Figure 4c,d, it indicates that the other Fe-containing phase of small quantity show in A3 and A4 alloys. Moreover, the slight

Cu/Mg-containing phases is residual in high-Cu level A4 alloy. Except the dissolution of Mg and Cu into the matrix, the eutectic Si happens evident spheroidization and dispersion, as shown in Si mapping images of Figure 4.

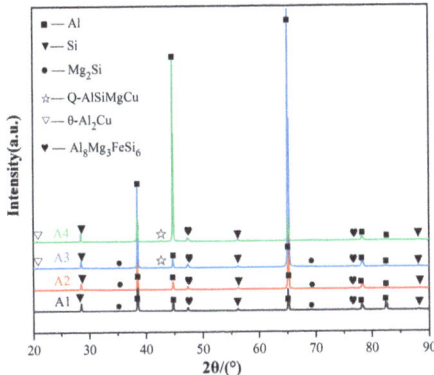

Figure 3. XRD image of as-cast alloys.

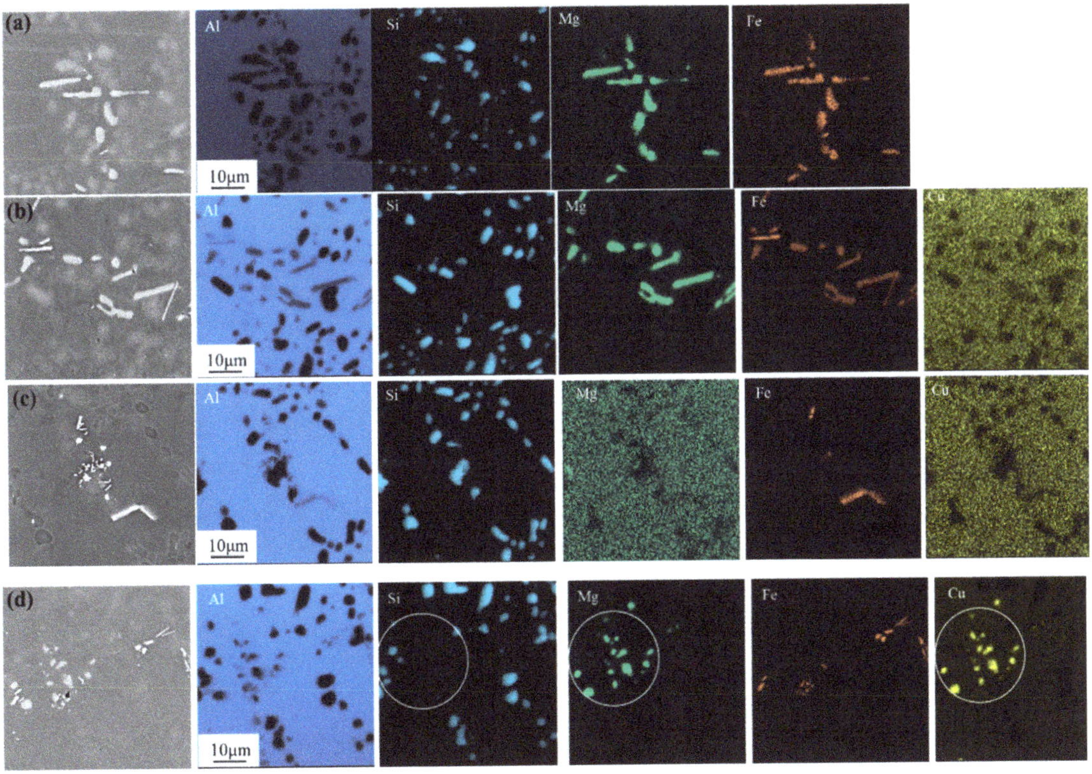

Figure 4. SEM-mapping images of the alloys after aging treatment (**a**) A1, (**b**) A2, (**c**) A3, (**d**) A4.

The TEM micrographs of the alloys after aging treatment are shown in Figure 5. The comparison of the TEM images of the A1–A4 alloys indicates that the nano-precipitates precipitated during aging process show a higher number density with increasing Cu level.

By contrast, the lamellar precipitates were observed in A4 alloy. Here, the lamellar precipitates are mainly Cu-containing phases, may be the sheet θ' platelets specifically [7,8]. The number density of precipitates is listed in Table 2. The number density of the precipitates was estimated by η = 3 N, where N is the precipitate cross-section in the image. The factor 3 comprises the three growth directions of the precipitates [13]. In Table 2, the increase in Cu content improves the number density of the precipitates in the aging-treated alloys.

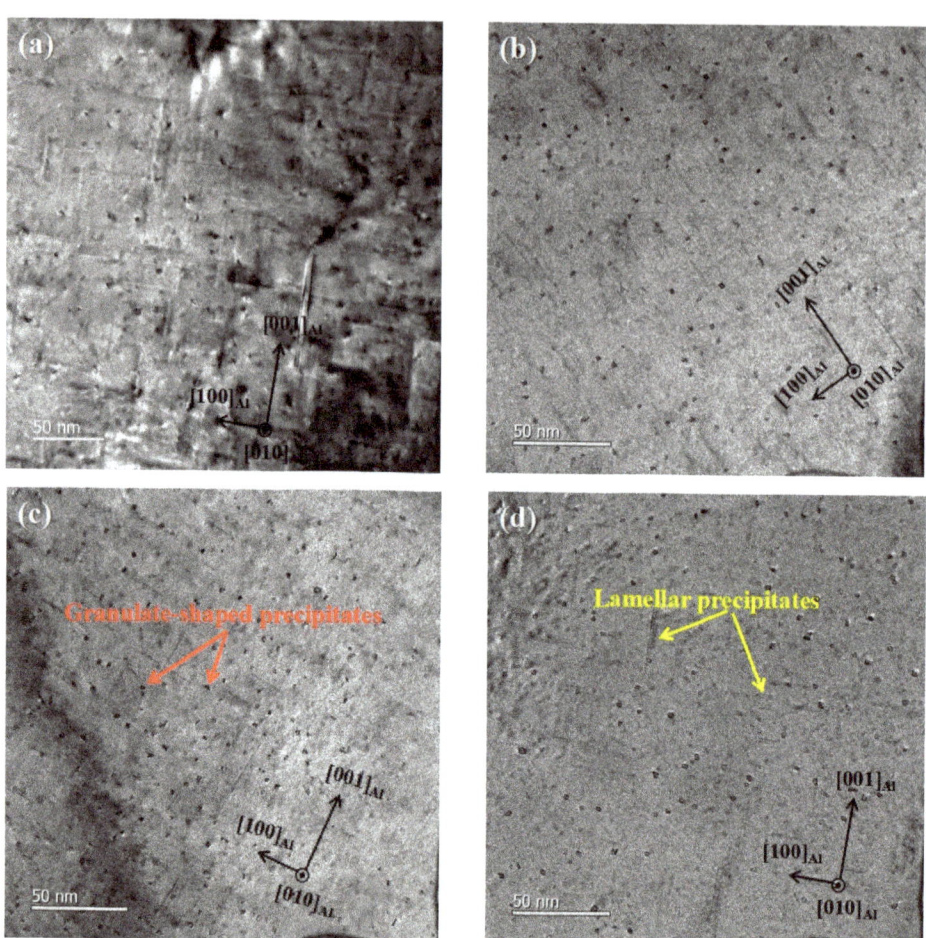

Figure 5. Bright-field TEM images of the alloys after aging treatment (a) A1, (b) A2, (c) A3, (d) A4.

Table 2. Number density of precipitate in aging-treated samples.

Alloy	n (Number Density/ × 10^{22} m^{-3})	l (Average Length/nm)	A_{cs} (Average Area of the Cross-Section/nm^2)
A1	6.56	22.41	16.6501
A2	6.98	22.85	17.8293
A3	8.01	24.52	19.6582
A4	8.21	27.59	22.3276

Figure 6 presents the TEM results for the precipitates in A1 alloy after aging treatment. This precipitate shows the monoclinic structure with lattice parameters of a = 1.51 and c = 0.67 nm (Figure 6b), in addition to the orientation relationships of (200)$_{precipitate}$//(301)$_{Al}$ (Figure 6b) and [010]$_{precipitate}$//[010]$_{Al}$ (Figure 6c,d). These results indicate that these precipitates are β″ phases [13,18], and no other precipitates are observed. This precipitate is regarded as the most common one in the aged Al alloy. HRTEM images (Figure 6b–d) show that the precipitate is coherent with the Al matrix.

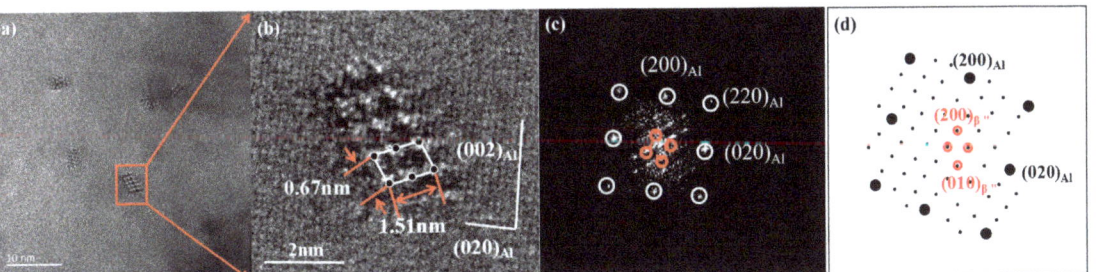

Figure 6. β″ phase in A1 alloy in the <010>Al axis after aging treatment: (**a**) TEM image, (**b**) HRTEM image, (**c**) corresponding FFT, (**d**) schematic pattern.

Figure 7 shows the TEM image and the corresponding FFT pattern of the precipitate in A2 alloy. The HRTEM image and FFT pattern indicate that β″ precipitates are in the aged A2 alloy. A closer check of A2 alloy reveals that the precipitation of another nanophase (Figure 8) in addition to the extensive β″ phase (Figure 7). This precipitate shows an angle of 120° between its a and b axes, and exhibits a typical dense stacked hexagonal lattice (HCP) crystal structure. The lattice parameter of this precipitate is a = 1.032 nm, which is obtained using an internal standard method. The precipitate interface was largely parallel to the three crystal faces of the Al matrix—(501)$_{Al}$, (103)$_{Al}$, and (506)$_{Al}$—but was mostly distributed along <510>Al with an orientation relationship of (2110)$_{precipitate}$//(501)$_{Al}$, [0001]$_{precipitate}$//[010]$_{Al}$. On the basis of previously reported literature data [19], this precipitate in A2 alloy is identified as granular Q′ phases. In general, the main precipitates in A2 alloy are the needle-like β″ phase and the granular Q′ phase. It can be seen from Figures 7 and 8 that the precipitates are coherent with the Al matrix.

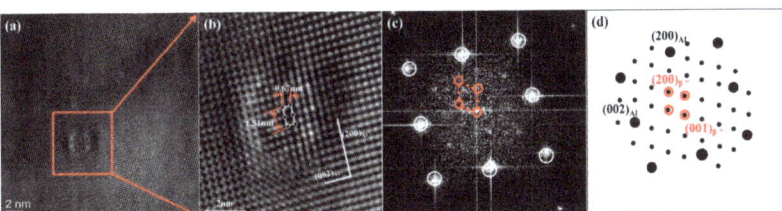

Figure 7. β″ phase in A2 alloy in the <010>Al axis after aging treatment: (**a**) TEM image, (**b**) HRTEM image, (**c**) corresponding FFT, (**d**) schematic pattern.

Figure 9 shows the TEM images and the corresponding FFT pattern of the granular precipitate in the aged A3 alloy. The Q′ nano-phase also can be observed, but no precipitate of other type was observed. Combined with the results in Figure 8, this result indicates that large quantity of Q′ phases can be identified in A3 alloy. The size of the Q′ phase is approximately the same as in A2 alloy, and the predominant precipitates in the aged A3 alloy are the granular Q′ phases, and it is coherent with the Al matrix. As can be seen in Figure 10, the precipitates in A4 alloy have a crystal structure, lattice parameter, and orientation relationship similar to those of the Q′ phase.

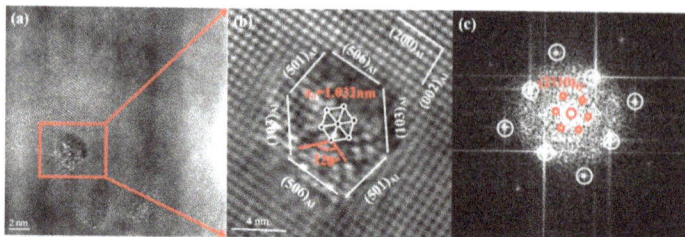

Figure 8. Q′ phase in A2 alloy in the <010>Al axis after aging treatment: (**a**) TEM image, (**b**) HRTEM image, (**c**) corresponding FFT.

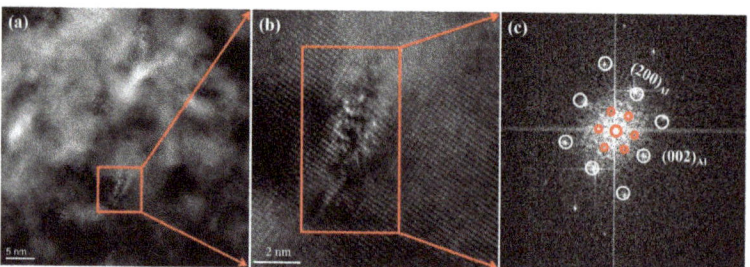

Figure 9. Q′ phase in A3 alloy in the <010>Al axis after aging treatment: (**a**) TEM image, (**b**) HRTEM image, (**c**) corresponding FFT.

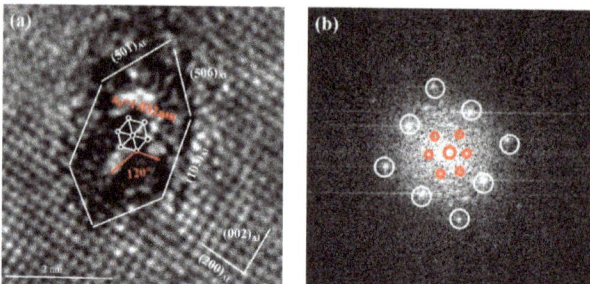

Figure 10. Q′ phase in A4 alloy after aging treatment along the <010>Al axis: (**a**) HRTEM image, (**b**) corresponding FFT.

Figure 11 shows the TEM images of the lamellar precipitates in A4 alloy, which indicates that these precipitates are distributed along {200} Al. A closer examination shows that the precipitates exhibit the crystal structure and lattice parameters: a = 0.404 nm, c = 0.58 nm, and an orientation relationship of $(200)_{precipitate}//(200)_{Al}$, $[010]_{precipitate}//[010]_{Al}$. Therefore, this nano-phase is identified as θ′ [7,20], and the predominant precipitates in the aged A4 alloy are Q′ and θ′. Additionally, normally, the reduction of interfacial energy causes the precipitates to be compact, while reduction of elastic energy leads to the formation of the plate shape. The ratio between the bulk elastic driving force and the interfacial energy is size dependent, and thus, the tendency towards plate formation depends on the precipitate size. The shape formation of smaller precipitates is mostly driven by interface reduction and therefore the precipitates tend to be more spherical when the interfacial energy is assumed to be isotropic. Combined with Figure 5, the precipitated phase was gradually changed from short rod to granulate to lamellar with increasing Cu content.

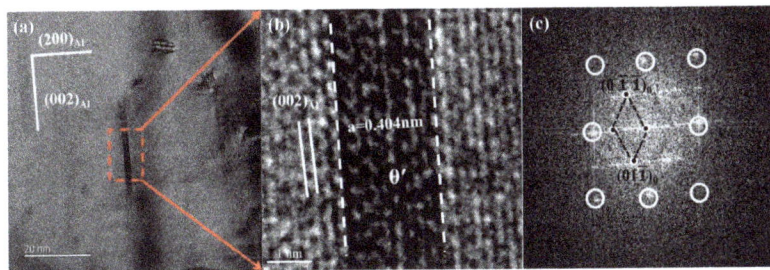

Figure 11. θ′ phase in A4 alloy after aging treatment, observed along the <010>Al axis: (a) TEM image, (b) HRTEM image, (c) corresponding FFT.

In summary, the types of aging-precipitate changes with the change of Cu level in alloys, which transforms from needle-like β″ in A1 alloy to granular Q′ and needle-like β″ in A2 alloy. A3 alloy mostly consists of granular Q′ phase. When Cu level increases to 2.08%, the Q′ and θ′ are the main precipitates in aged A4 alloy.

3.3. Tensile Properties

The ultimate tensile strength (UTS), 0.2% yield strength (YS), and the elongation to fracture of the solution- and aging-treated alloys are listed in Table 3. The strength of the studied alloys increases, and the elongation slightly decreases with increasing of Cu level. Figure 12 presents the engineering stress–strain curves of the alloys under quenching and aging conditions. Under solution-treated state, the YS increases from 161 to 221 MPa and the UTS increases from 275 to 363 MPa when Cu content increases from 0–2%. Meanwhile, the solution-treated alloys of A1–A4 show the high elongations of 16–18%. After aging treatment, the YS and UTS markedly improve in A1–A4 alloys. With increasing Cu content, the YS increases from 264 to 351 MPa, UTS increases from 322 to 442 MPa, and the elongation decreases from 10 to 8.4%.

Table 3. Tensile properties of the designed alloys under different conditions.

Alloy	Quenching State			Aging Treatment		
	UTS (MPa)	YS (MPa)	A_{25} (%)	UTS (MPa)	YS (MPa)	A_{25} (%)
A1	275 ± 5.2	161 ± 4.2	18.5 ± 1.2	322 ± 4.2	264 ± 4.6	10 ± 1.1
A2	335 ± 5.3	194 ± 4.4	17.1 ± 1.3	343 ± 5.7	299 ± 3.5	8.7 ± 0.8
A3	347 ± 4.5	201 ± 3.3	16.4 ± 0.9	394 ± 6.1	321 ± 4.1	8.6 ± 0.7
A4	363 ± 4.2	221 ± 4.5	16.1 ± 1.6	442 ± 5.3	351 ± 4.9	8.4 ± 0.6

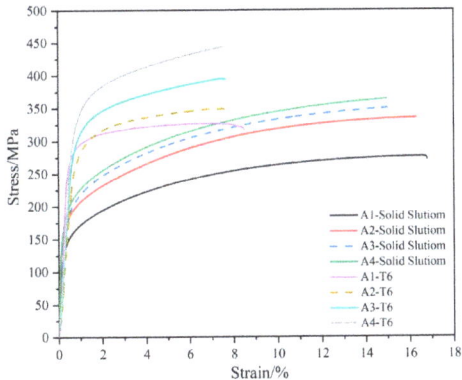

Figure 12. Engineering stress–strain curves of the alloys after solid solution and aging treatment.

Comparing the tensile properties of the alloys treated under different conditions, the change in strength of the alloys with different Cu contents can attribute to solid-solution and aging strengthening. In solution-treated samples, the increase of Cu content results in a higher solution content, which shows a higher solution strengthen effect. The YS and UTS of solution-treated samples gradually increase when Cu content increases. After aging treatment, the mechanical properties of the alloys markedly increase; the aging-treatment is more effective than solution treatment on influence of strength in the Al-Si-Mg-xCu cast alloy. Moreover, it indicates that the elongations of aging-treatment samples decrease due to the inverted relationship between strength and ductility.

Notably, the yield strengths of the Al-Si-Mg-xCu alloys markedly improve after aging treatment. The improvement in yield strength can be attributed to the nano-precipitates of the β″, Q′, and θ′ phases. The contribution of the precipitates to yield strength can be calculated by the Ashby–Orowan equation [13]:

$$\sigma_D = \frac{0.84 MGb}{2\pi(1-v)^{1/2}\lambda} \ln\frac{r}{b}, \qquad (1)$$

where M is the Taylor factor, $M = 3.1$, G and b were the shear modulus (2.65×10^{10} N/m^2) and the Burgers vector of dislocations in the Al matrix (2.84×10^{-10} m), and v is the Poisson's ratio for Al (0.33). The interspacing of the precipitates λ depends on the radius r and volume fraction V_f of the precipitates, as follows:

$$\lambda = r\left(\frac{2\pi}{3v_f}\right)^{\frac{1}{2}}, \qquad (2)$$

Volume fraction (V_f) of precipitates:

$$V_f = nlA_{cs}, \qquad (3)$$

where n is the number density of precipitates, l is the average length of dispersoids, and A_{cs} is the average area of the cross-section of precipitates. According to Equations (2)–(4), the increase of yield strength caused by precipitates is calculated, which is listed in Table 4.

Table 4. Differences in the yield strength of the samples between quenching and aging and the contribution to yield strength from precipitates calculated by the Orowan mechanism.

Alloy	Yield Strength after Quenching (MPa)	Yield Strength after Aging (MPa)	Improvement in Yield Strength (MPa) (by Experiment)	Precipitates Contribution to Yield Strength (MPa) (by Orowan Mechanism)
A1	161 ± 4.2	264 ± 4.6	103	119.5
A2	194 ± 4.4	299 ± 3.5	105	125.0
A3	201 ± 3.3	321 ± 4.1	120	128.1
A4	221 ± 4.5	351 ± 4.9	130	135.8

As can be seen from Table 4, the calculated data agree well with the improved experimental data of yield strength after aging process. However, the measured increase is a little bit lower than the calculated increase in yield strength contributed by the precipitates. This difference can be explained by the decrease of solute strengthening, since the solute concentration in solid solution decreases during aging process [13,21,22].

According to the Orowan bypass mechanism, the yield strength increment ($\Delta\sigma_s$) shows the relationship with the f and r (Equation (4)), where α is a constant for the material, f

is the second-phase particle (aging precipitates) volume percentage, and r is the average radius of the second phase particle (aging precipitates) [13,22]:

$$\Delta\sigma_s \propto \alpha \cdot f^{1/2} \cdot r^{-1} \tag{4}$$

This relationship was used to analyze variation tendency of yield strength increment with the precipitate size, as shown in Figure 13. It indicates that the increasing effect of the aging precipitates on the yield strength is proportional to $f^{1/2} \cdot r^{-1}$. Therefore, the increment of the yield strength is increased with the increase of $f^{1/2} \cdot r^{-1}$ in Figure 13, which is consistent with the Orowan mechanism.

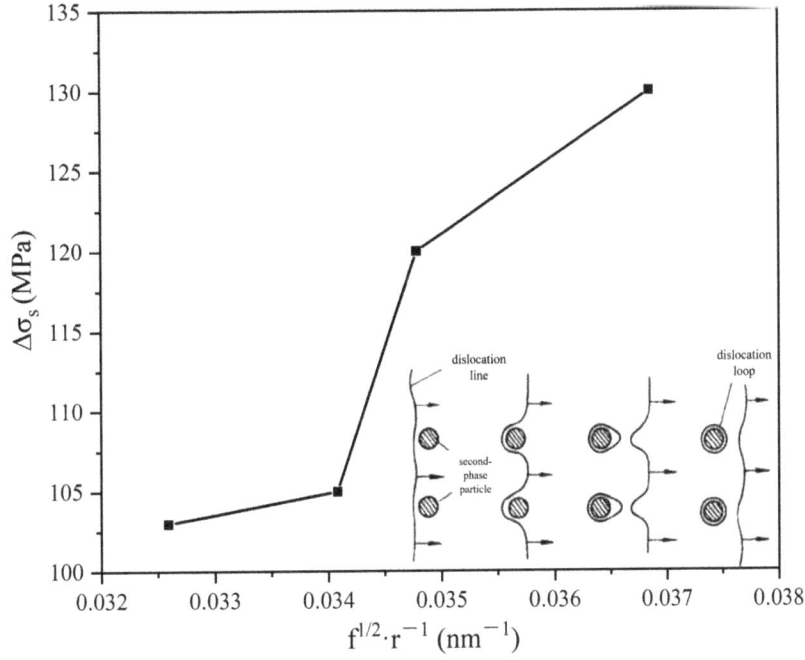

Figure 13. The variation tendency of yield strength increments with the $f^{1/2} \cdot r^{-1}$.

Figure 14 shows the fracture surfaces of the alloys after T6 treatment. The casting defect does not be observed in the fracture surfaces. The fracture surfaces of the four alloys are occupied by dimples formed by spheroidizing and dispersive Si particles. These dispersive Si particles can result in the fine and homogeneous dimples during one-axis loading process, which is important for high ductility of Al-Si-Mg-Cu cast alloy. In the loading-bearing process, most granular Si particles were pulled out in dimples, and dimples formed in the Al matrix, represented by the yellow arrows in Figure 14. Moreover, the fracture surfaces in T6-treated samples do not show the residual Cu- and Fe-containing phases, avoiding adverse effect on the ductility by coarse intermetallic. Therefore, the elongations of aging-treated samples are higher than 8%, which can meet the engineering application requirements of Al-Si-Cu-Mg casting alloy.

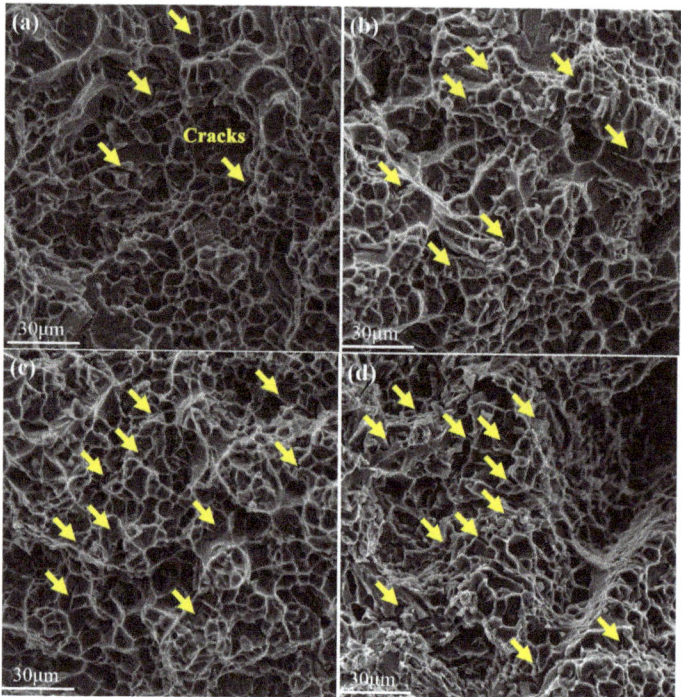

Figure 14. Fracture surfaces of the alloys after T6 treatment: (**a**) A1, (**b**) A2, (**c**) A3, (**d**) A4.

4. Conclusions

The microstructures and mechanical properties of Al-9Si-0.5Mg alloys with Cu addition were investigated. The following conclusions are drawn from this study.

(1) With the Cu level increasing and Cu/Mg ratio changing, the mechanical properties including yield and ultimate tensile strengths improve after solution and aging treatments in A1–A4 alloys, the increase of Cu content results in higher solution and aging strengthen effects and affects the aging precipitates. The aging-treated Al-9Si-0.5Mg-2Cu alloy shows a better strength and ductility: yield strength 351 MPa, ultimate tensile strength 442 Mpa, and elongation 8.4%.

(2) With the Cu level increasing, the types of main precipitates in the aging-treated samples changes from the needle-like β″ phase in base alloy to the β″ and granular Q′ phase in the 0.9%Cu alloy. When Cu content further increases to 1.5% and 2.1%, the types of main precipitates show the Q′ phase in the 1.5%Cu alloy, and the Q′ and θ′ phases in the 2.1%Cu alloy. After aging, the increase in Cu level leads to the increase in the volume fraction, number density, and average cross-sectional area of the precipitates.

Author Contributions: Conceptualization, D.W. and H.N.; Methodology, P.Z., D.W. and X.Z.; Validation, X.Z., X.L. and B.Z.; Formal Analysis, R.W. and B.Z.; Investigation, H.N., H.Z. and R.W.; Resources, D.W., B.Z. and R.W.; Data curation, P.Z., B.Z. and X.Z.; Writing—Original draft preparation, P.Z. and D.W.; Writing—Review & Editing, P.Z., D.W., X.Z. and R.W.; Visualization, X.L., H.Z. and B.Z.; Supervision, D.W., H.N. and H.Z.; Project administration, X.L., H.N. and H.Z.; Funding acquisition, X.L., D.W. and H.N. All authors have read and agreed to the published version of the manuscript.

Funding: This research was funded by the National Natural Science Foundation of China (Grants no. 52004168), The Research Fund for International Senior Scientists (Grants no. 52150710544), the

National Natural Science Foundation of China (Grants nos. U1864209 and 51771066), the Aluminum-based Transportation Lightweighting Technology Demonstration Project (Grants no. 2021SFGC1001), and the National Foreign Expert Project 2022 (Grants no. G2022014146L). And The APC was funded by The Research Fund for International Senior Scientists (Grants no. 52150710544).

Institutional Review Board Statement: Not applicable.

Informed Consent Statement: Not applicable.

Data Availability Statement: The raw/processed data required to reproduce these findings cannot be shared at this time as the data also forms part of an ongoing study.

Conflicts of Interest: The authors declare no conflict of interest.

References

1. Liu, T.; Pei, Z.-R. Characterization of nanostructures in a high pressure die cast Al-Si-Cu alloy. *Acta Mater.* **2022**, *224*, 117500. [CrossRef]
2. Ma, S.; Wang, Y.; Wang, X. The in-situ formation of Al3Ti reinforcing particulates in an Al-7wt%Si alloy and their effects on mechanical properties. *J. Alloys Compd.* **2019**, *792*, 365–374. [CrossRef]
3. Dong, X.; Zhu, X.; Ji, S. Effect of super vacuum assisted high pressure die casting on the repeatability of mechanical properties of Al-Si-Mg-Mn die-cast alloys. *J. Mater. Process. Technol.* **2019**, *266*, 105–113. [CrossRef]
4. Dong, X.; Amirkhanlou, S.; Ji, S. Formation of strength platform in cast Al-Si-Mg-Cu alloys. *Sci. Rep.* **2019**, *9*, 1–11. [CrossRef]
5. Zhang, G.-W.; Wang, Z.-J. Age hardening of Al-7Si-0.5Mg alloy: Role of Si size and distribution. *J. Alloys Compd.* **2023**, *933*, 167797. [CrossRef]
6. Mohamed, A.M.A.; Samuel, E.; Samuel, A.M. Effect of Intermetallics and Tramp Elements on Porosity Formation and Hardness of Al-Si-Mg and Al-Si-Cu-Mg Alloys. *Int. J. Met. Cast.* **2022**, 1–18. [CrossRef]
7. Chen, B.; Dong, L. The Effect of Cu Addition on the Precipitation Sequence in the Al-Si-Mg-Cr Alloy. *Materials* **2022**, *22*, 8221. [CrossRef]
8. Li, H.; Guo, S.L.; Du, P.; Liu, S. Effect of Cu Content on Microstructure and Properties of Al-Mg-Si Alloy. *Phys. Eng. Met. Mater.* **2019**, *217*, 143–151.
9. Zhu, X.Z.; Ji, S. Improvement in as-cast strength of high pressure die-cast Al-Si-Cu-Mg alloys by synergistic effect of Q-Al5Cu2Mg8Si6 and θ-Al2Cu phases. *Mater. Sci. Eng. A* **2021**, *802*, 140612. [CrossRef]
10. Zuo, L.; Ye, B. Effect of Q-Al5Cu2Mg8Si6 phase on mechanical properties of Al-Si-Cu-Mg alloy at elevated temperature. *Mater. Sci. Eng. A* **2017**, *693*, 26–32. [CrossRef]
11. Jonas, K.S.; Calin, D.M.; Randi, H. The effect of low Cu additions on precipitate crystal structures in overaged Al-Mg-Si(-Cu) alloys. *Mater. Charact.* **2020**, *160*, 110087.
12. Shang, X.J.; Liu, Q.; Xu, P. Effects of copper and rare Earth elements on properties of aluminum electrical round bars. *Trans. Nonferrous Met. Soc. China* **2018**, *8*, 16–19.
13. Wang, Y.-F.; Lu, Y.-L. Characterization and strengthening effects of different precipitates in Al-7Si-Mg alloy. *J. Alloys Compd.* **2021**, *885*, 161028. [CrossRef]
14. Zhou, P.; Wang, D.; Liu, S.; Wang, R.; Zhang, H.; Li, X.; Nagaumi, H. New Strategy to Improve the Mechanical Properties in Cast Al-Mg-Fe Alloys by the Formation of Al-AlFe Eutectic. *Adv. Eng. Mater.* **2021**, *23*, 2001460. [CrossRef]
15. Yan, P.; Mao, W. Microstructural evolution, segregation and fracture behavior of A390 alloy prepared by combined Rheo-HPDC processing and Sr-modifier. *J. Alloys Compd.* **2020**, *835*, 155297. [CrossRef]
16. Pramod, S.; Ravikirana, A.; Prasadarao, B.; Murty, S. Effect of Sc addition and T6 aging treatment on the microstructure modification and mechanical properties of A356 alloy. *Mater. Sci. Eng. A* **2016**, *674*, 438–450. [CrossRef]
17. Shabani, M.; Mazahery, A. Prediction of mechanical properties of cast A356 alloy as a function of microstructure and cooling rate. *Arch. Metall. Mater.* **2011**, *56*, 671–675. [CrossRef]
18. Shishido, H.; Aruga, Y.; Murata, Y.; Marioara, C.; Engler, O. Evaluation of precipitates and clusters during artificial aging of two model Al–Mg–Si alloys with different Mg/Si ratios. *J. Alloys Compd.* **2022**, *927*, 166978. [CrossRef]
19. Farkoosh, A.; Pekguleryuz, M. Enhanced mechanical properties of an Al–Si–Cu–Mg alloy at 300 C: Effects of Mg and the Q-precipitate phase. *Mater. Sci. Eng. A* **2015**, *621*, 277–286. [CrossRef]
20. Li, Y.; Brusethaug, S.; Olsen, A. Influence of Cu on the mechanical properties and precipitation behavior of AlSi7Mg0.5 alloy during aging treatment. *Scr. Mater.* **2006**, *54*, 99–103. [CrossRef]
21. Zhang, X.; Liu, F. Microstructural evolution and strengthening mechanism of an Al-Si-Mg alloy processed by high-pressure torsion with different heat treatments. *Mater. Sci. Eng. A* **2020**, *794*, 139932. [CrossRef]
22. Lei, W. *Mechanical Properties of Materials*, 3rd ed.; China Machine Press: Beijing, China, 2014; p. 76.

Disclaimer/Publisher's Note: The statements, opinions and data contained in all publications are solely those of the individual author(s) and contributor(s) and not of MDPI and/or the editor(s). MDPI and/or the editor(s) disclaim responsibility for any injury to people or property resulting from any ideas, methods, instructions or products referred to in the content.

Article

Research on the 2A11 Aluminum Alloy Sheet Cyclic Tension–Compression Test and Its Application in a Mixed Hardening Model

Guang Chen, Changcai Zhao *, Haiwei Shi, Qingxing Zhu, Guoyi Shen, Zheng Liu, Chenyang Wang and Duan Chen

Key Laboratory of Advanced Forging and Stamping Technology and Science of Ministry of Education, Yanshan University, Qinhuangdao 066004, China
* Correspondence: zhao1964@ysu.edu.cn; Tel.: +86-185-3351-1399

Abstract: The increasing application of aluminum alloy, in combination with the growth in the complexity of components, provides new challenges for the numerical modeling of sheet materials. The material elastic-plasticity constitutive model is the most important factor affecting the accuracy of finite element simulation. The mixed hardening constitutive model can more accurately represent the real hardening characteristics of the material plastic deformation process, and the accuracy of the material property-related parameters in the constitutive model directly affects the accuracy of finite element simulation. Based on the Hill48 anisotropic yield criterion, combined with the Voce isotropic hardening model and the Armstrong–Frederic nonlinear kinematic hardening model, a mixed hardening constitutive model that considers material anisotropy and the Bauschinger effect was established. Analysis of the tension–compression experiment on the sheet using finite element method. Using the finite element model, the optimum geometry of the tension–compression experiment sample was determined. The cyclic deformation stress–strain curve of the 2A11 aluminum alloy sheet was obtained by a cyclic tensile–compression test, and the material characteristic parameters in the mixed hardening model were accurately determined. The reliability and accuracy of the established constitutive model of anisotropic mixed hardening materials were verified by the finite element simulation and by testing the cyclic tensile–compression problem, the springback problem, and the sheet in bending, unloading, and reverse bending problems. The tensile–compression experiment is an effective method to directly and accurately obtain the characteristic parameters of constitutive model materials.

Keywords: 2A11 aluminum alloy plate; anisotropic; Bauschinger effect; mixed hardening; cyclic tension–compression experiment

1. Introduction

The use of stamping to form parts is common in various fields, such as metal material processing, the aerospace industry, the automobile industry, and scientific research [1,2]. The finite element numerical simulation technology is an effective means of shortening the stamping die design cycle, achieving process optimization, and improving the quality of stamping parts. In the stamping process with cyclic loading characteristics, the selection of an elastic–plastic constitutive model and related hardening behavior are of great significance in predicting the actual forming process [3]. A kinematic hardening model and a mixed hardening model can accurately represent the true hardening characteristics during plastic deformation. The accuracy of material characteristic parameters in the constitutive model directly affects the accuracy of the finite element simulation [4].

The elastic–plastic constitutive model of materials includes three components: yield criterion, the flow rule, and a hardening model. In simulating stamping and forming, the commonly used hardening models can be divided into isotropic hardening models, kinematic hardening models, and mixed hardening models. In the isotropic hardening models, the subsequent yield surface B only changes in size and position relative to the initial yield surface A, as shown in Figure 1a. The typical isotropic hardening models are the Mises model and the Hill model, which are simple and easy to program. However, they can only describe the similar changes in the yield surface under a single strain path and they cannot describe some changes in material properties (such as the Bauschinger effect and the cross effect) when the strain path changes [5]. In the kinematic hardening models, the size of the subsequent yield surface remains unchanged only when the position changes, as shown in Figure 1b. The Ziegler model and the Armstrong–Frederic (A–F) model are widely used. Ziegler [6] proposed linear kinematic hardening based on the proportional relationship between the back stress increment and the strain increment. The Armstrong–Frederic nonlinear kinematic hardening model introduced a dynamic recovery item with decreasing memory for the deformation path, eliminated the defects of linear kinematic hardening, and better described the Bauschinger effect, which is the research foundation for the nonlinear kinematic hardening model [7]. The kinematic hardening model avoids the isotropic hardening model's drawback of being unable to describe the Bauschinger effect, but it cannot describe the expansion of the yield surface during deformation [8,9].

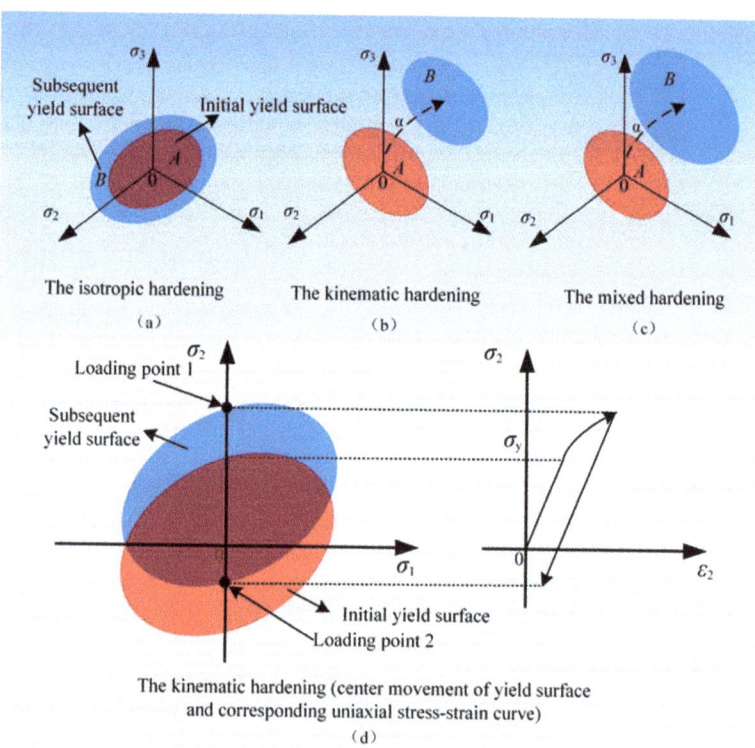

Figure 1. Schematic diagram of yield surface variation in the classic hardening theories: (a) isotropic hardening model; (b) kinematic hardening model; (c) combined hardening model; (d) kinematic hardening (center movement of yield surface and corresponding uniaxial stress–strain curve).

The above two hardening models only describe part of the hardening behavior of materials. In plastic deformation, the yield surface of most materials undergoes both size and position changes. Therefore, when describing the hardening behavior of actual metal materials, the above two hardening models are often combined and a mixed hardening model is used, as shown in Figure 1c. Han [10] and Li Qun et al. [11] established a mixed hardening model based on the Voce isotropic hardening model and the A–F kinematic hardening model, introduced the hardening model into an equivalent drawbead resistance model, proposed an equivalent drawbead model that considered the Bauschinger effect, and verified the accuracy of the model via experiments. With the development of ABAQUS software (version 6.14, Dassault Systemes Simulia Corp., Providence, RI, USA) for finite element analysis, some parameters related to the material properties required by the kinematic hardening model and the mixed hardening model can be directly input into the software without sophisticated secondary development.

ABAQUS (version 6.14, Dassault Systemes Simulia Corp., Providence, RI, USA) provides linear and nonlinear kinematic hardening models to simulate the cyclic loading of metals. The linear kinematic model has a constant hardening modulus, which is suitable for analyzing hardening behavior with an approximately constant hardening rate. The non linear kinematic hardening model defines the kinematic hardening part as an incremental combination of a pure motion term (the linear Ziegler hardening rule) and a relaxation term (the recall term), so that nonlinearity is introduced to the kinematic hardening part. At the same time, in ABAQUS/Explicit, the yield stress ratio R_{ij} of the input plate can be used with the Hill48 yield surface [12]. ABAQUS (version 6.14, Dassault Systemes Simulia Corp., Providence, RI, USA) provides three methods—assigning parameter input, assigning semi-cyclic tension-compression experiment data, and assigning sheet cyclic tension-compression experiment data to define the kinematic hardening part. The kinematic hardening material characteristic parameters obtained from the cyclic stress–strain curve of the sheet can most accurately reflect the hardening behavior of the plate under cyclic loading. The cyclic sheet tension–compression test is required to obtain the cyclic sheet tensile–compression stress–strain curve.

The buckling tendency of a thin sheet under compression is very large, and it is difficult to obtain a large compression strain. In order to obtain the tensile and compressive properties of a thin sheet under large strain, scholars have proposed various methods to suppress the buckling of the thin plate in the compression process. Boger et al. [13] used a solid plate to clamp the two sides of the thin sheet sample and applied normal pressure via a hydraulic clamping system to constrain the buckling of the thin sheet during compression. However, the thickness of the thin plate changes during the tensile–compression process, and there will inevitably be a gap between the chuck of the tensile machine and the anti-buckling fixture, resulting in excessive test error. Kuwabara et al. [14] designed a device with two pairs of comb teeth to reduce areas that are not clamped. However, the comb-tooth area is prone to bending, and the comb-tooth device is expensive. Yoshida [15] designed a special device to combine multiple samples for the tensile–compression test to overcome the defect of instability of a single sample. It measured the strain in the test process, but could not avoid the compression instability under strain. On the basis of a wedge-shaped unit designed by Cheng et al. [16], Cao et al. [17] designed an anti-buckling wedge fixture using transparent materials, which could be used to measure the whole optical strain in the deformation region of the sample by an optical strain-measurement method. Although this method can obtain tensile–compression strain under a large strain, the sample is prone to lateral instability. For the compression tests, Kurukuri [18] and Abedini et al. [19] prepared bonded sheet laminates to overcome any buckling during the tests. Due to the action of the glue between the plates, the test results had a large error, and the preparation of the sample was very complicated. In summary, in order to make the cyclic tensile–compression test of a thin sheet perform smoothly, the following three problems must be solved:

- designing a set of a reasonable thin-plate in-plane normal constraint device of thin sheets that can not only prevent the instability of the sample, but also minimize the increase in the axial compressive capacity caused by the friction force of the normal constraint;
- determining the best sample geometry that can minimize the effect of in-plane buckling and improve the accuracy of stress measurement;

- selecting a high-precision optical strain-measuring instrument in order to accurately measure the deformation of the thin sheet sample gauge.

Based on the above research, this paper aimed to establish an A–F nonlinear kinematic hardening constitutive model according to the Hill48 anisotropic yield criterion. Based on the Hill48 anisotropic yield criterion, the Voce isotropic hardening model, and the A–F nonlinear kinematic hardening model, a constitutive model inclusive of anisotropy and mixed hardening was established. A set of sheet tension–compression buckling-restrained fixtures was designed, and the shape parameters that affect stress-measurement error and inhibit in-plane buckling were determined. The optimal shape of the sample used in the in-plane compression test was determined to accurately obtain the material properties-related parameters of the kinematic hardening model and the mixed hardening model. ABAQUS (version 6.14, Dassault Systemes Simulia Corp., Providence, RI, USA) finite element software was used to analyze the applicability of the constitutive model for the cyclic tensile–compression problem of the sheets and the problem of the plate after bending, unloading, and reverse bending. The accuracy and reliability of the constitutive model were verified by experiments. This provides a reliable research method for studying the deformation behavior of sheet metal under complex loading conditions with cyclic loading characteristics.

2. Description of the Constitutive Model

2.1. Establishment and Parameter Determination of the Nonlinear Kinematic Hardening Constitutive Model

Most of the sheets used in stamping had anisotropy and a high material hardening rate, so the Mises yield criterion and the isotropic hardening model could not truly reflect the plastic behavior of the sheets during deformation. The Hill48 yield criterion considers the anisotropic characteristics of the material and considers that the contribution of stress in each direction of the sheet to the plastic yield is different, which information can be used for the plastic description of the sheet-forming process.

Assuming that the thickness anisotropy index r is constant during the plastic deformation process, if the sheet metal conforms to the flow rule of the total strain theory, then the r value can be obtained by measuring the strains in the width direction (ε_w) and thickness direction (ε_t) using a single tensile test. Specifically, the r value is expressed as follows:

$$r = \frac{\varepsilon_w}{\varepsilon_t} \tag{1}$$

The expressions for the ratio of six anisotropic yield stresses [12]—R_{11}, R_{22}, R_{33}, R_{12}, R_{13} and R_{23}—can be derived by combining the Hill48 anisotropic yield conditions.

For the anisotropic behavior and the Bauschinger effect exhibited by the material during plastic deformation, the kinematic hardening model provides a simple explanation that the yield surface of the material only moves as a rigid body and does not rotate in the stress space during deformation, and the back stress represents the center of the plastic yield surface in the stress space.

The yield surface function of kinematic hardening materials is generally expressed as follows:

$$\Phi = F(\sigma_{ij} - \alpha_{ij}) - \sigma_Y = 0 \tag{2}$$

where σ_Y is the initial yield stress and α_{ij} is the back stress.

The back stress represents the movement of the center of the yield surface in the stress space, which plays a crucial role in the kinematic hardening model and in yield surface evolution. Its value is related to the material hardening characteristics and deformation history. As shown in Figure 1d, the material is subjected to unidirectional elastic–plastic loading along the direction of σ_2, and the stress increases from $\sigma_2 = 0$ to $\sigma_2 = \sigma_y$. During the loading process, when the deformation state of the material changes from elastic deformation to plastic deformation, the center of the yield surface begins to move. When the stress in the σ_2 direction is loaded to loading point 1, unloading and reverse loading are implemented to deform the material and the material stress reaches the loading point 2 to produce reverse plastic yield. It is clear from Figure 1d that the yield stress of the material under reverse loading is smaller than the initial yield stress σ_y.

The A–F nonlinear kinematic hardening model has been extensively utilized to study the cyclic plastic behavior of materials. This model contains a linear hardening term and a dynamic restoration term. The evolution equation can be expressed as follows:

$$d\alpha_{ij} = \frac{2}{3}C d\varepsilon_{ij}^p - \gamma \alpha_{ij} d\bar{\varepsilon}^p \qquad (3)$$

where C and γ are material parameters; α_{ij} is the back stress component; $d\varepsilon_{ij}^p$ is the increment in the plastic strain; $d\bar{\varepsilon}^p$ is the equivalent plastic strain increment [11]; and $d\bar{\varepsilon}^p = \sqrt{\frac{d\varepsilon_{ij}^p}{d\varepsilon_{ij}^p}}$.

When the material is subjected to uniaxial tensile loading, $d\bar{\varepsilon}^p = d\varepsilon^p$. Therefore, it can be determined that

$$d\alpha_1 = \frac{2}{3}C d\varepsilon_1^p - \gamma \alpha_1 d\varepsilon_1^p \qquad (4)$$

The above equation can be simplified as follows:

$$\frac{d\alpha_1}{\frac{2}{3}C - \gamma \alpha_1} = d\varepsilon_1^p \qquad (5)$$

Integrating the above first-order differential equation, we obtain

$$\alpha_1 = \frac{2}{3}\frac{C}{\gamma} + \left(\alpha_1^0 - \frac{2}{3}\frac{C}{\gamma}\right) e^{-\gamma(\varepsilon_1^p - \varepsilon_{1,0}^p)} \qquad (6)$$

where α_1^0 is the initial value of the back stress and $\varepsilon_{1,0}^p$ is the initial value of the plastic strain.

The initial conditions are $\alpha_1^0 = \varepsilon_{1,0}^p = 0$. Using these initial values, the back stress equation can be obtained as follows:

$$\alpha_1 = \frac{2}{3}\frac{C}{\gamma}\left(1 - e^{-\gamma \varepsilon_1^p}\right) \qquad (7)$$

According to the above equation, the parameters C and γ can be determined by the nonlinear fitting of the experimentally acquired plastic strain data and the real stress obtained from the uniaxial tensile test of the sheet sample. The six anisotropic parameters (R_{11}, R_{22}, R_{33}, R_{12}, R_{13}, and R_{23}) and the kinematic hardening parameters (C and γ) were input into the ABAQUS (version 6.14, Dassault Systemes Simulia Corp., Providence, RI, USA) material model library to obtain the material characteristic parameters related to the follow-up hardening constitutive model, based on the Hill48 yield criterion.

2.2. Establishment of the Mixed Hardening Constitutive Model

According to the Hill48 anisotropic yield criterion, combined with the Voce isotropic hardening model and the A–F nonlinear kinematic hardening model, a constitutive model that considers anisotropy and mixed hardening was established. The isotropic hardening part adopted the Voce nonlinear isotropic hardening criterion, and the equation is as follows:

$$\bar{\sigma} = \sigma_0 + Q\left(1 - e^{-b\varepsilon_1^p}\right) \qquad (8)$$

where ε_1^p is the plastic strain, Q and b are the parameters of the isotropic hardening materials, and σ_0 is the initial yield stress.

Under the uniaxial stress state, combined with the Equation (9), the material flow stress can be expressed as follows:

$$\sigma = \sigma_0 + Q\left(1 - e^{-b\varepsilon_1^p}\right) + \frac{2}{3}\frac{C}{\gamma}\left(1 - e^{-\gamma\varepsilon_1^p}\right) \tag{9}$$

In order to determine the nonlinear relationship between flow stress σ and plastic strain ε_1^p, it is necessary to determine the four material characteristic parameters Q, b, C, and γ. These material characteristic parameters can be obtained by the cyclic tensile–compression stress–strain curve obtained by the sheet cyclic tensile–compression test.

2.3. Material and Experimental Procedure

The 2A11 aluminum alloy plate with a thickness of 0.6 mm was selected as the research object. 2A11 aluminum alloy is a hard aluminum alloy that is widely used in the aerospace industry, the transportation industry, and other fields. The sheet sample was cut with a line cutting machine, according to the China National Standard "Tensile testing method for metal materials at room temperature" (GB/T 228-2002). The uniaxial tensile test of the original sheet was conducted on the InspektTable-100 material universal testing machine (Huibo, Germany). The standard size of the sheet sample used in the unidirectional tensile test was 50 mm. Three groups of unidirectional tensile samples were cut along the directions of 0°, 45°, and 90° with respect to the rolling direction. The strain rate during the tensile test was 0.0013/s. To obtain the anisotropy coefficient, the axial strain, and the transverse strain during the deformation process of the sample were recorded by an online strain-measurement system based on digital image correlation (DIC). The testing machine and the DIC online strain-measurement system are shown in Figure 2a, and the size of the uniaxial tensile sample is shown in Figure 2b.

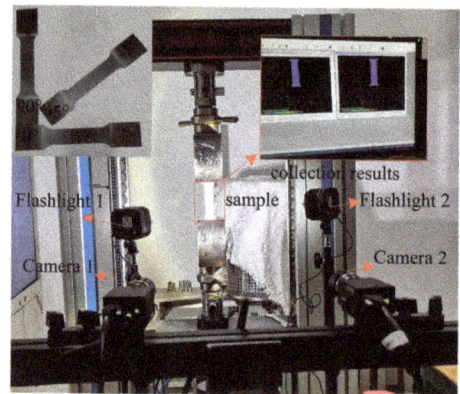

(a) Tensile test device

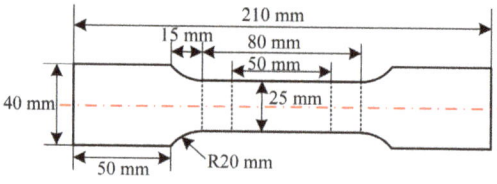

(b) Schematic diagram of tensile sample

Figure 2. Uniaxial tensile test diagram of 2A11 aluminum alloy sheet: (**a**) test drawing machine diagram; (**b**) uniaxial tensile sample.

The engineering stress–strain curve of the large sample was obtained by the uniaxial tensile test, and the real stress–strain data were obtained according to the conversion formula, as shown in Figure 3a. In the uniaxial tensile process of the sample in three directions, the strains in the width and thickness directions were measured to obtain the r value, and the results are shown in Figure 3b. According to Equation (9), the initial yield point in the real stress–strain curve of 2A11 aluminum alloy sheet was taken as the starting point of the back stress parameter fitting curve for obtaining the relevant parameters. The parameters of each material model are listed in Tables 1 and 2.

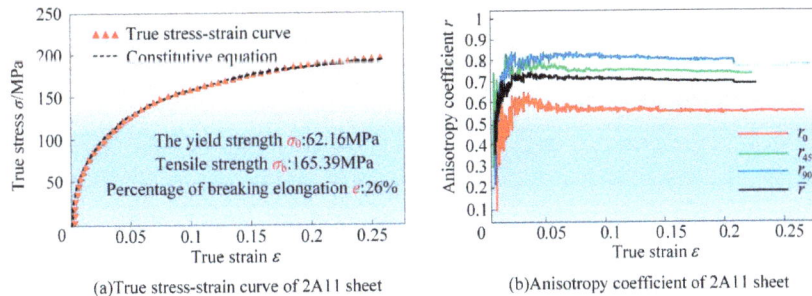

(a) True stress–strain curve of 2A11 sheet

(b) Anisotropy coefficient of 2A11 sheet

Figure 3. (a) True stress–strain curve of 2A11 aluminum alloy sheet; (b) anisotropy coefficient of 2A11 aluminum alloy sheet.

Table 1. Simulation of material parameters required.

Material	Yield Strength (MPa)	Poisson's Ratio	Young's Modulus (GPa)	C/MPa	γ
2A11	62.16	0.31	68.016	1258	152

Table 2. Ratio of anisotropic yield stress of materials.

Material	R11	R22	R33	R12	R13	R23
2A11	1	1.09	0.96	1.05	1	1

3. Cyclic Sheet Tension-Compression Test and Sample Shape Optimization

3.1. Shape Optimization of Cyclic Sheet Tension-Compression Sample

The cyclic tensile–compression test of aluminum bars is relatively easy to perform and there are corresponding standards to follow [20,21]. However, an aluminum alloy sheet is prone to instability during compression, and the optimum sample shape to minimize the stress-measurement error and suppress the in-plane buckling of the sample has not yet been defined [22]. Therefore, this study used the finite element method (FEM) to identify the shape parameters that had an effect on the stress-measurement error and the shape parameters that had an effect on the suppression of in-plane buckling to determine the optimum shape of the sample for use in the cyclic tension-compression tests.

Figure 4 represents the shape parameters of the sheet tensile–compression sample evaluated in this study. The width and length of the parallel section are W and L, respectively; the radius of the fillet of the transition section between the parallel section and the clamping end is R; the width of the clamping end is B and the length of the clamping end was determined by the clamping head size of the test machine and was set at 30 mm.

When a sheet tensile–compression sample was subjected to in-plane compression, the deformation within the parallel section was not uniform due to the transition section constraining the deformation near the ends of the parallel section, resulting in inconsistent compressive stresses. Therefore, the sample shape had to be optimized to improve the accuracy of the stress measurement. In addition, in order to retard the onset of in-plane buckling and to allow greater compression to be applied, shape parameters that helped to suppress in-plane buckling had to be defined to optimize the cyclic drawing of the sheet samples.

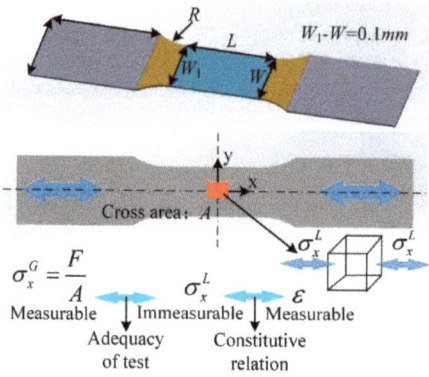

Figure 4. Geometrical parameters of the sample for in-plane compression test and schematic illustration of the difference between mean stress $\sigma_x^{(G)}$ and local stress $\sigma_x^{(L)}$.

In the cyclic sheet tensile–compression test, the average stress $\sigma_x^{(G)}$ can be found from the compression force F and the area of the central section of the sample derived from the constant volume criterion. However, the local stress $\sigma_x^{(L)}$ at the center of the sample in the experiment could not be obtained in the sample and could be extracted in the post-processing results of the FEM. The smaller the difference between the absolute value of the local stress $\sigma_x^{(L)}$ and the mean stress $\sigma_x^{(G)}$ in the central part of the sample, the higher the accuracy of the stress measurement. Therefore, the FEM was used to analytically determine the sample shape that minimizes the difference between the local stress $\sigma_x^{(L)}$ and the mean stress $\sigma_x^{(G)}$.

The relative deviation of $\sigma_x^{(L)}$ from $\sigma_x^{(G)}$ is given by the following equation:

$$\tau_m = \frac{\sigma_x^{(G)} - \left|\sigma_x^{(L)}\right|}{\left|\sigma_x^{(L)}\right|} \times 100\% \tag{10}$$

The smaller the τ_m sample shape, the greater the accuracy of the stress measurement.

The deformation was not uniform in the parallel part of the sample, due to cyclic tension–compression. To clarify the sample shape parameters that inhibit surface buckling using the FEM, the width difference $W-W_1 = 0.05$ mm between the two ends of the parallel section was set as the initial unevenness to perform the buckling analysis. Under compression, the parallel section of the sample will produce shear stress $\sigma_{xy}^{(L)}$. The ratio of shear stress $\sigma_{xy}^{(L)}$ in the central part of the sample to local stress $\sigma_x^{(L)}$ was used to determine the buckling of the sample. The determination formula of in-plane buckling is as follows:

$$\tau_s = \frac{\left|\sigma_{xy}^{(L)}\right|}{\left|\sigma_x^{(L)}\right|} \times 100\% \tag{11}$$

The smaller the τ_s, the less the possibility of buckling during compression.

3.2. FEM Model for Sample Shape Optimization

To obtain the optimum sample shape for the cyclic sheet tensile–compression test, a finite element model for the cyclic sheet tensile–compression test was established, as shown in Figure 5a. The aspect ratio of the deformation zone to the width of the clamping end B and the corner radius R of the corner transition zone had a great influence on the accuracy of the stress measurement and the compression limit, so a variety of dimensional parameters were set for the FEM analysis, as shown in Figure 5b. The clamping plates on both sides of the sample deformation zone were always clamped with a clamping force of 940 N. The clamping plates on both sides of the clamping end were bound together with the clamping end of the sample. The friction coefficient between the cyclic tension–compression sample and the clamping plates was set to 0.084. The strain rate during the cyclic tension-compression test was 0.0013/s. In the finite element model for the cyclic sheet tensile–compression test, the cyclic sheet tensile–compression sample was set up as a deformed body and all the remaining components were defined as rigid bodies. The sample was set up with five integration points in the thickness direction by applying a four-node curved thin-shell or thick-shell reduction integral, an S4R cell with finite film strain, and an hourglass control. The element size of the sample was set to 0.5 mm to ensure that no distortion occurred in the process of compression, so as to obtain accurate stress calculation results.

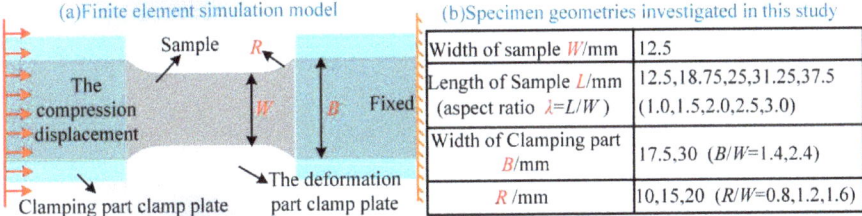

Figure 5. (a) Finite element simulation model; (b) sample geometries investigated in this study.

3.3. Sample Shape Optimization Simulation Results

3.3.1. Effect of Sample Deformation Zone Aspect Ratio λ and Clamping End Width B

During the compression of the cyclic sheet tensile–compression sample, $B/W = 1.4$, $R/W = 1.2$; $B/W = 2.4$, $R/W = 1.2$ were set to investigate the effect of the aspect ratio λ of the deformation zone of the sample and the width B of the clamping end. From the simulation results, as shown in Figure 6a,b, when the width B of the sample clamping end and the radius R of the fillet area were certain, the stress-measurement accuracy gradually increased with the increase in the sample deformation zone aspect ratio λ. The stress-measurement accuracy decreased due to the increase in the length of the sample deformation zone and the influence of the corner area on the measurement accuracy became smaller when the width of the clamping end B was too large. Buckling was more likely to occur when the sample deformation zone was too long, as shown in Figure 6c,d. The wider the sample clamping end, the more serious the deformation of the rounded area as the transition area, which exerted more influence on measurement accuracy. From the simulation results and the above analysis, it can be seen that the longer the deformation zone of the sample, the higher the accuracy of the stress measurement, but buckling was also more likely to occur; therefore, the best aspect ratio of the deformation zone of the sample was moderately chosen as $\lambda = 2$. The deformation of the fillet area of the sample directly affects the accuracy of the stress measurement, so it was necessary to investigate the influence of the shape and size of the fillet area on the accuracy of the stress measurement.

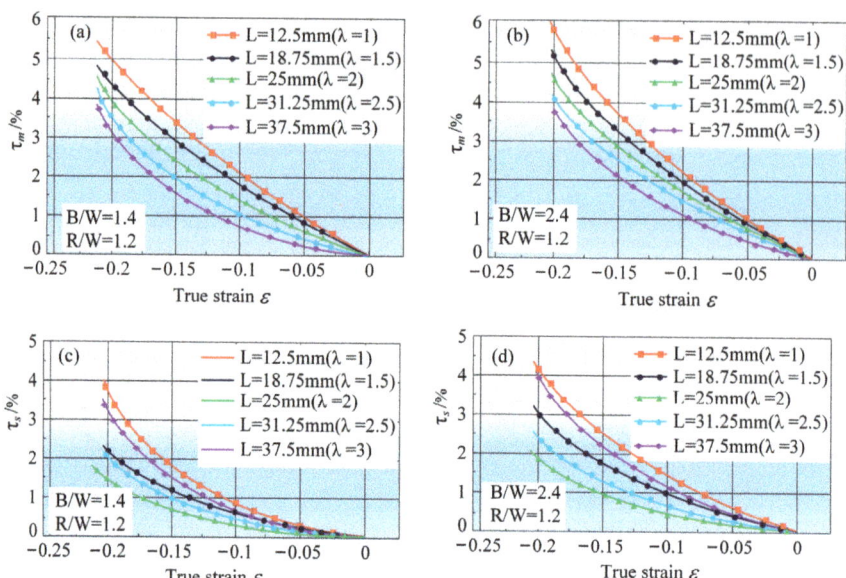

Figure 6. Effect of aspect ratio λ on the variation of τ_m and τ_s with true strain: (**a**) B/W = 1.4; (**b**) R/W = 1.2; (**c**) B/W = 2.4; (**d**) R/W = 1.2.

During the compression of the cyclic sheet tensile–compression sample, B/W = 1.4, R/W = 1.2; B/W = 2.4, R/W = 1.2 were set to investigate the effect of the aspect ratio λ of the deformation zone of the sample and the width B of the clamping end. From the simulation results, as shown in Figure 6a,b, when the width B of the sample clamping end and the radius R of the fillet area were certain, the stress-measurement accuracy gradually increased with the increase of the sample deformation zone aspect ratio λ.

3.3.2. Effect of Fillet Radius R in the Fillet Area of the Sample

Applying B/W = 1.4, λ = 2, different sizes of fillet radius R were set to analyse the effect on stress-measurement accuracy and buckling. The simulation results are shown in Figure 7. An increase in the fillet radius improved the accuracy of stress prediction and it was less likely that buckling occurred. The reason for this was that the large fillet radius area weakened the restraint on both ends of the parallel section, so the parallel section deformed more uniformly and was less likely to buckle.

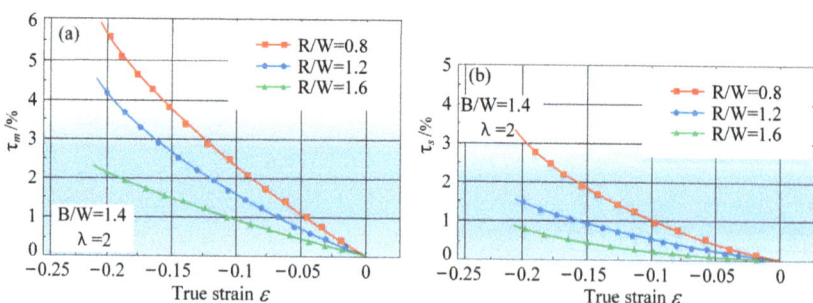

Figure 7. Effect of fillet radius R on (**a**) τ_m and (**b**) τ_s.

3.3.3. Influence of n-Values of the Sample

With sheets of different materials, n-values (the strain hardening exponent in Swift's equation) vary widely, and it is known from studies related to buckling formation that the larger the n-value of the sheet, the less likely it is to buckle [23]. The uniaixal tensile curve of 2A11 aluminum alloy sheet was fitted and its n value was obtained as 0.3468. The n-values of 0.2, 0.3468 and 0.5 were set, and the parallel section aspect ratio $\lambda = 2$ and $B/W = 1.4$ were used for FEM analysis. The results are shown in the Figure 8. With the increase in n-value, the stress-measurement accuracy was higher and the tensile samples were less likely to buckle during the compression process. In summary, the sheets with large n-values had a higher stress-measurement accuracy in sheet drawing and in the cyclic sheet tensile–compression test, due to stable deformation and the sample was less prone to buckling; the plates with large n-values had better forming performance in stamping and forming with cyclic loading characteristics.

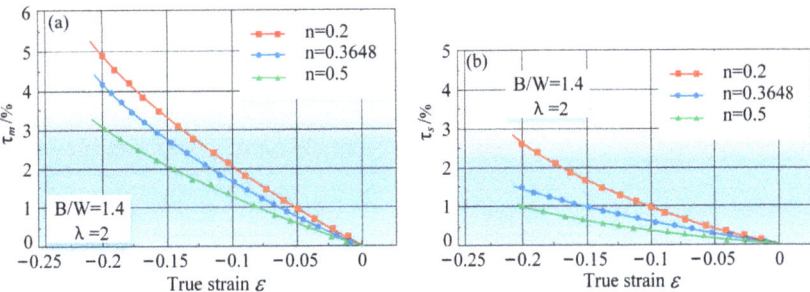

Figure 8. (**a**) Effect of n-values on τ_m; (**b**) Effect of n-values on τ_s.

Based on the above analysis of the FEM results for the optimization of the cyclic sheet tensile–compression sample shape, it can be seen that the sample shape with a moderate aspect ratio of parallel sections and larger radius of fillet was selected for higher accuracy of stress measurement and less susceptibility to buckling. The final determination of the shape and size of the cyclic sheet tensile–compression sample is shown in Figure 9b.

3.4. The Cyclic Sheet Tensile–Compression Test

To obtain the cyclic tensile–compression stress–strain curves of the studied sheets, an anti-buckling fixture was designed, as shown in Figure 9a. The cyclic tensile–compression test of the studied sheets used the specimen geometry of Figure 9b. The anti-buckling fixture was made of transparent acrylic sheets on both sides as raw material, and optical strain-measurement equipment was used to accurately measure the strain change of the sample. The transparent acrylic sheet did not affect the DIC camera's shooting, as shown in Figure 9c. The sheet had a change in thickness during the tensile–compression process, so the disc spring placed between the bolt and the clamping plate ensured that the clamping plate was always clamping elastically during the clamping process.

The relative sliding between the tensile–compression sample and the clamping plate generated friction, and the direction of the friction force was opposite to the direction of the movement of the tester chuck, making the load measured by the tester large. Assuming that the friction force was uniformly distributed on the contact surface, the Coulomb friction formula was applied to calculate the magnitude of the friction force during the test, and the friction force was removed from the measured test data to eliminate the error caused by friction on the load measurement. In order to determine the friction coefficient between the sheet and the clamping plate, the friction coefficient between the 2A11 aluminum alloy sheet and the acrylic plate was tested using a friction and wear tester produced by the Center for Tribology (CETR) in the United States The friction coefficient, using transparent silicone oil as a lubricant, was 0.084. The anti-buckling fixture was used to perform cyclic tensile–compression tests on sheet metal on the InspektTable-100 material universal testing machine, and the strain changes during sample deformation were recorded by the DIC online strain-measurement system.

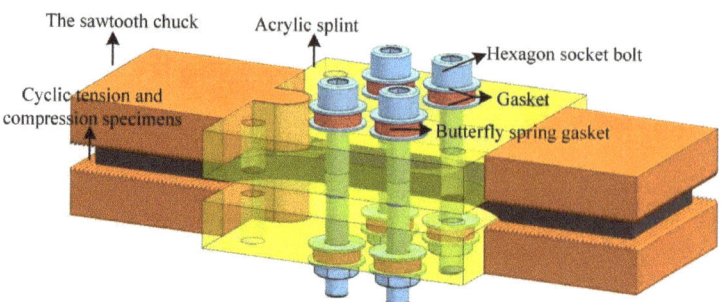

(a) schematic plot of cyclic tension and compression test

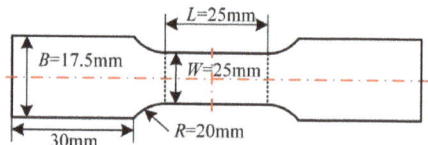

(b) schematic plot of tensile compression sample

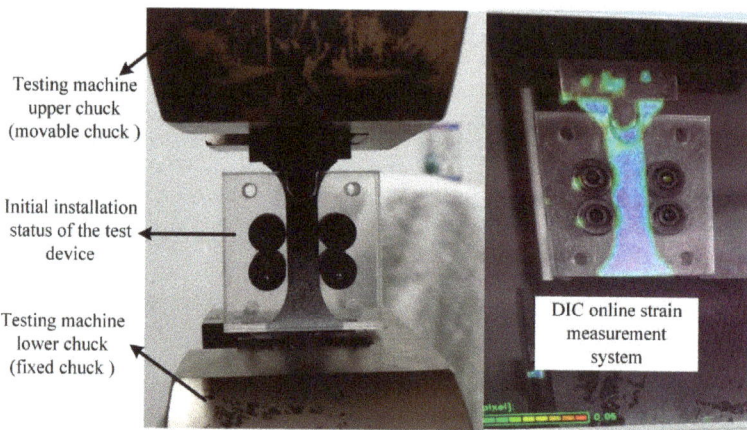

(c) cyclic tension and compression test device

Figure 9. Cyclic tensile–compression test: (**a**) schematic diagram of cyclic tension and compression test; (**b**) schematic drawing of used tensile–compression specimen; (**c**) cyclic tension and compression test device.

The disc spring gasket was of type A (GB/T1972-2005), with dimensions of outer diameter $\varphi 10$ mm, inner diameter $\varphi 5.2$ mm, thickness 0.5 mm, initial height 0.75 mm, and compressible amount 0.25 mm. The disc springs buckled in two groups of the same specification, each group made up of three stacked disc springs, and the combined buckling disc spring group was compressed by 0.75 mm, as shown in Figure 10a, while the stiffness curve of the single group of disc springs was measured by the InspektTable-100 material universal testing machine (Huibo, Germany). In the cyclic sheet tensil–compression test, the pre-compression of the disc spring group was 0.3 mm, and the total clamping force of the four groups of disc springs varied with the thickness of the sample, as shown in Figure 10b. The cyclic stress–strain curve for the selected specimen of Figure 9b, after eliminating the effect of fixture friction, is shown in Figure 10c. Similar data were obtained by many cyclic tensile–compression tests, and the test results were reliable.

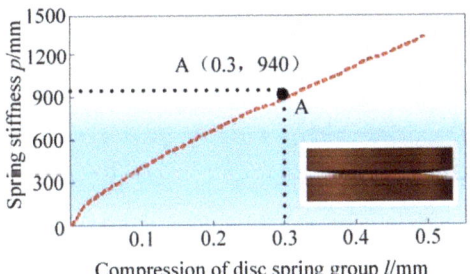

(a) Stiffness curve of disc spring group

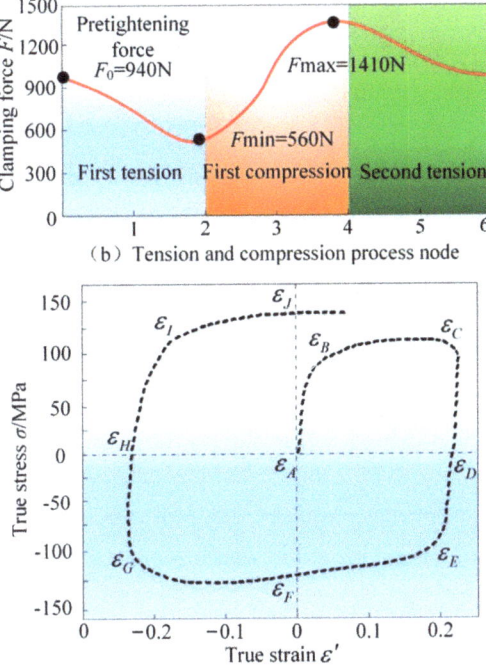

(b) Tension and compression process node

(c) Cyclic tension and compression stress-strain curve for the selected specimen of Fig. 9(b)

Figure 10. (a) Stiffness curve of butterfly spring group/mm; (b) tension and compression process node; (c) cyclic tension and compression stress–strain curve for the selected specimen of Figure 9b.

3.5. Determination of the Mixed Hardening Constitutive Model Parameter

The flow stress and the equivalent plastic strain in the constitutive model were nonlinearly mapped, and there were many parameters. In order to obtain each parameter of the constitutive model accurately, the solution was based on the cyclic sheet tensile–

compression stress–strain curve. The material flow stress in the uniaxial stress state in uniaxial tension for the mixed hardening constitutive model can be expressed as follows:

$$\sigma = \sigma_0 + Q\left(1 - e^{-b\varepsilon_1^p}\right) + \frac{2}{3}\frac{C}{\gamma}\left(1 - e^{-\gamma \varepsilon_1^p}\right) \qquad (12)$$

When the uniaxial tension reached $\sigma = \sigma_c$ and the plastic pre-strain was equal to ε_D during reverse loading and unloading, the flow stress was calculated as follows:

$$\sigma = -\sigma_0 - Q\left(1 - e^{-b\varepsilon_1^p}\right) - \frac{2}{3}\frac{C}{\gamma}\left(1 - e^{-\gamma \varepsilon_1^p}\right)\left(1 - (2 - e^{-\gamma \varepsilon_D})e^{\gamma(\varepsilon_D - \varepsilon_1^p)}\right) \qquad (13)$$

When it is reverse loading to ε_G, it is reverse loading again. At this time, the plastic strain is ε_H, and the flow stress is as follows:

$$\sigma = \sigma_0 + Q\left(1 - e^{-b\varepsilon_1^p}\right) + \frac{2}{3}\frac{C}{\gamma}\left(1 - \left(2 - (2 - e^{-\gamma \varepsilon_D})e^{\gamma(\varepsilon_H - \varepsilon_D)}e^{\gamma(2\varepsilon_D - \varepsilon_H - \varepsilon_1^p)}\right)\right) \qquad (14)$$

Let $A = \sigma_0 + Q + \frac{C}{\gamma}$, $B = -Q$, $C = b$, and $E = \gamma$, and through the summary Equations (12)–(14), it can be obtained that the general stress-strain equation in the process of the uniaxial cyclic tension-compression loading process is as follows:

$$\sigma = A + Be^{-C\varepsilon_1^p} \pm De^{-E\varepsilon_1^p} \qquad (15)$$

where, A, B, C, and E are material constants that can be obtained by fitting the cyclic tensile–compression stress–strain curve derived from the cyclic sheet tensile–compression test, and D is a constant related to the direction of the pre-strain, with a plus sign for the forward direction and a minus sign for the reverse direction.

According to the cyclic tensile–compression stress–strain curves derived from the cyclic tensile–compression tests, four material characteristic parameters—Q, b, C, and γ—in the mixed hardening model can be obtained by fitting. The goodness of fit R^2 was 0.99 and the fitting accuracy was high. Q and b are the parameters related to isotropic hardening; C and γ are the parameters related to kinematic hardening, as shown in Table 3.

Table 3. Parameters obtained in the combined hardening model.

Material	Yield Strength (MPa)	Q/MPa	B	C/MPa	γ	R^2
2A11	62.16	65.5	37	1453	168	0.99

4. Reliability Analysis of the Constitutive Model

4.1. The Application of the Constitutive Model in the Cyclic Sheet Tensile–Compression Problem

The material characteristic parameters of the A–F nonlinear kinematic hardening constitutive model, based on the Hill48 anisotropic yield criterion, and the material characteristic parameters of the mixed hardening constitutive model, based on the Hill48 anisotropic yield criterion, the Voce isotropic hardening model, and A–F nonlinear kinematic hardening model, were input into the cyclic sheet tensile–compression simulation model. The simulation results of the two constitutive models were compared with the experimental results, as shown in Figure 11.

In the initial tensile stage, the simulated results of the two constitutive models were in general agreement with the experimental results. When the sheet was subjected to tensile deformation unloading, and reverse compression, the simulated results deviated from the experimental results, with an average deviation of 7.4% for the simulated results of the kinematic hardening model and 2.1% for the simulated results of mixed hardening model. When the sheet was stretched again, the average deviation of the simulated results of mixed hardening model was 1.3%, while the average deviation of the simulated results of kinematic hardening model was 11.7%.

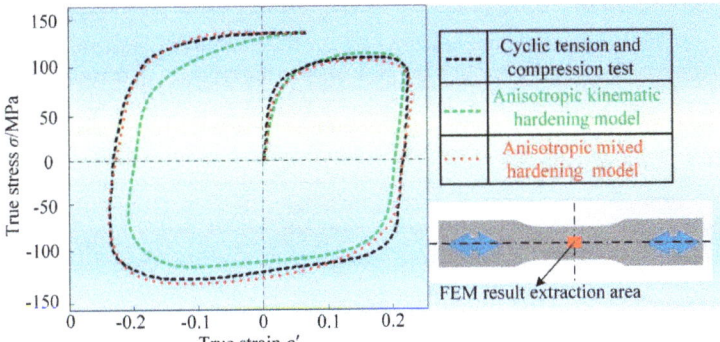

Figure 11. Fitting and calibration of constitutive elastic–plastic material model parameters and comparison of simulation and experimental results.

4.2. The Application of the Constitutive Model in the Continuous Bending Problem

The kinematic hardening model and the mixed hardening constitutive model were applied to the simulation of continuous bending, straightening, and reverse bending deformation. The FEM model was established for simulation analysis, as shown in Figure 12. The continuous bending model structural parameters were as follows: the transitions rounding of the upper and lower die were 7.34 mm and 8 mm, respectively; the radii of the semicircular rounding of the upper and lower die were 8.66 mm and 8 mm, respectively; the 2A11aluminum alloy sheet with a thickness of 0.6 mm was chosen as the object of study, and the sheet's length, width, and thickness were 100 mm, 10 mm, and 0.6 mm, respectively. After the upper and lower die closed the die, a gap of 0.06 mm was retained. A 4-node reduced integration S4R unit was used to divide the sheet, and five integration points were set in the sheet thickness direction. In order to ensure the smooth progress of the continuous bending test, no fracture occurred in the sample, Coulomb friction was used between the contact surface of the upper and lower die and the sheet, and the friction factor was set to 0.096. The front end of the sheet was bound to the pulling plate and the displacement constraint was used.

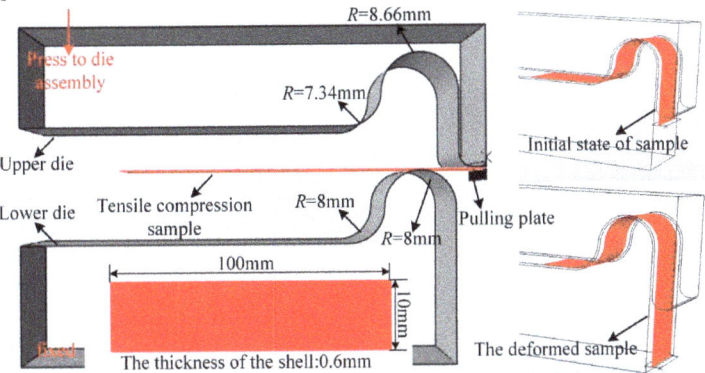

Figure 12. Finite element simulation model.

The material characteristic parameters of the A–F nonlinear kinematic hardening constitutive model, based on the Hill48 anisotropic yield criterion, and the material characteristic parameters of the mixed hardening constitutive model, based on the Hill48 anisotropic yield criterion, the Voce isotropic hardening model, and the A–F nonlinear kinematic hardening model, were input into the sheet continuous-bending model and compared with the sheet continuous-bending test for analysis.

The continuous-bending test die of the sheet is shown in Figure 13. Figure 13a,b show the upper and lower die of the continuous bending test die, respectively, and the size is the same as the size of the die in the simulation model. The upper and lower dies were fixed by bolts, and the upper chuck of the InspektTable-100 material universal testing machine clamped the clamping end of the upper die, and the lower chuck of the universal testing machine clamped the lower end of the continuous bending sample, as shown in Figure 13c. In the continuous-bending test, the specimen was easy to break because of the large bending deformation. In order to ensure the smooth progress of continuous-bending test without fracturing the sample, an oil-based molybdenum disulfide lubrication was applied between the continuous-bending sample and the mold, and the friction coefficient was 0.096. During the test, the upper chuck was fixed and the lower chuck was moved down at a constant speed of 50 mm/min. The state of the continuously bent sample after the die assembly and after experiencing continuous bending is shown in Figure 13d.

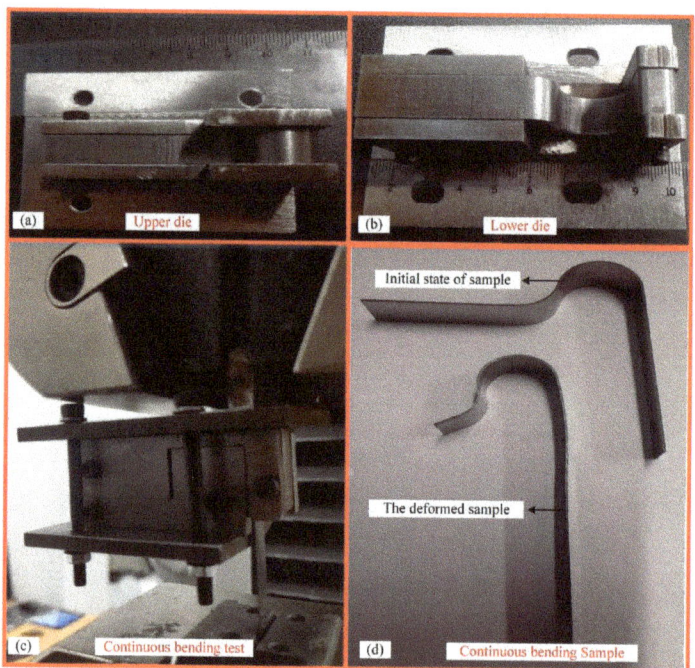

Figure 13. (**a**) Upper die; (**b**) Lower die; (**c**) Continuous bending test die; (**d**) Continuous bending Sample.

Figure 14 shows the results of the comparison between the lower chuck tension during the continuous-bending test and the pulling plate tension in the simulation results. The computational error of the mixed hardening model is smaller than that of the kinematic hardening model and is closer to the test value. The wall thickness distribution of the sample after continuous bending was measured with a micrometer and compared with the simulation results of the two constitutive models in the same state. The wall-thickness distribution of the samples in the simulation results of the kinematic hardening model and the mixed hardening model was consistent with the trend of the experimental results, and the simulation results of the mixed hardening model were closer to the experimental results, with little deviation. It was demonstrated that the material characteristic parameters of the mixed hardening constitutive model, based on the Hill48 anisotropic yield criterion, the Voce isotropic hardening model, and the A–F nonlinear kinematic hardening model, can be used to study the continuous-bending deformation behavior of the sheet under a complex loading condition, and its calculation results are reliable.

	Simulation results of kinematic hardening model	Simulation results of mixed hardening model	Results of experimental	Error(%)	
				The kinematic hardening model	The mixed hardening model
Tensile force F/kN	5.416	5.842	5.768	6.1	1.3

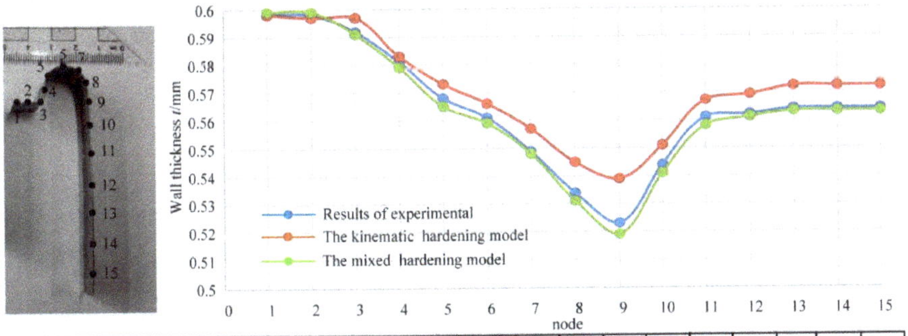

Results of experimental t/mm	0.598	0.598	0.592	0.581	0.568	0.561	0.549	0.534	0.523	0.544	0.561	0.562	0.564	0.564	0.564
The kinematic hardening model t/mm	0.598	0.597	0.597	0.583	0.573	0.566	0.557	0.545	0.539	0.551	0.567	0.569	0.572	0.572	0.572
The mixed hardening model t/mm	0.599	0.599	0.591	0.579	0.565	0.559	0.548	0.531	0.519	0.541	0.558	0.561	0.563	0.563	0.563

Figure 14. Comparison between FEM results and experimental results.

4.3. Application of Constitutive Model in the Springback Problem

A three-point bending FEM model was created, as shown in Figure 15, to test the correctness of the two constitutive models used to simulate the springback issue. The three-point bending test model's indenter diameter was 20 mm, the support point diameter was 30 mm, the distance between the two support points was 100 mm, and the three-point bending sample's length, width, and thickness were 120 mm, 20 mm, and 0.6 mm, respectively. The sheet material was divided into five integration points in the direction of the sheet thickness using a 4-node reduced integration S4R cell. Coulomb friction was used between the contact surfaces, and the friction factor was set to 0.1. Two support points were fixed, and the indenter was constrained by displacement. The material characteristic parameters of the A–F nonlinear kinematic hardening constitutive model, based on the Hill48 anisotropic yield criterion, and the material characteristic parameters of the mixed hardening constitutive model, based on the Hill48 anisotropic yield criterion, the Voce isotropic hardening model, and the A–F nonlinear kinematic hardening model, were input into the sheet three-point bending model and compared with the three-point bending test for analysis.

In sheet metal forming, the prediction of springback is important to show the desired final geometrical quality of the parts. Figure 15 presents the final result after springback was obtained from the kinematic hardening model, the mixed hardening model, and the experimental results. The results of the mixed hardening model matched the experimental data with superior accuracy than the results of the kinematic model, according on comparisons of the final shapes after the anticipated springback.. This proved that the mixed hardening model implemented in ABAQUS (version 6.14, Dassault Systemes Simulia Corp., Providence, RI, USA) can be applied for an accurate prediction of springback in complex industrial sheet metal forming operations.

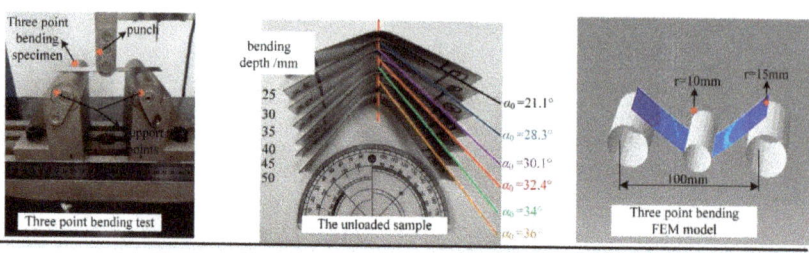

Rolling reduction /mm	Bending angle before unloading a_1 /°	Bending angle after unloading a_2 /°	Springback angle a_0 /°	The kinematic hardening model		The mixed hardening model	
				Springback angle a_0 /°	Error(%)	Springback angle a_0 /°	Error(%)
25	141.5	120.4	21.1	20.7	-1.89	20.8	-1.42
30	134.6	106.3	28.3	27.6	-2.47	27.9	-1.41
35	128.6	98.5	30.1	28.9	-3.98	30.6	1.66
40	123.6	91.2	32.4	31.1	-4.01	33.1	2.16
45	119.5	85.5	34	31.9	-6.17	35.2	3.52
50	115.9	79.9	36	33.1	-8.05	37.7	4.72

Figure 15. Comparison of the FEM results and the experimental results.

5. Conclusions

(1) The FEM analysis of the in-plane compression sample quantitatively specified the shape parameters that minimize the measurement error of compressive stress and help to suppress in-plane buckling. A transparent anti-buckling fixture was designed to accurately measure the strain variation of the cyclic tensile–compression sample using optical strain-measurement equipment.

(2) An anisotropic mixed hardening constitutive model based on the Hill48 anisotropic yield criterion, the Voce isotropic hardening model, and the A–F nonlinear kinematic hardening model was established. The cyclic deformation stress–strain curves of 2A11 aluminum alloy sheet were obtained by cyclic sheet tensile–compression tests, and the material characteristic parameters in the mixed hardening model were accurately determined. The obtained material characteristic parameters were directly input into the Abaqus simulation software, eliminating the need for tedious secondary development.

(3) The reliability and accuracy of the established constitutive model for anisotropic mixed hardening materials were verified through finite element simulations and tests of the aluminum alloy sheet cyclic tension-compression problem, the springback problem, and the sheet in bending, unloading, and reverse bending problems. It provided a reliable research tool for predicting the deformation behavior of the sheet under complex loading conditions with cyclic loading characteristics.

Author Contributions: Conceptualization, G.C.; methodology, G.C., Q.Z. and C.W.; software, G.C. and H.S.; validation, C.Z., Q.Z. and D.C.; formal analysis, C.Z.; investigation, C.Z., H.S., Z.L., C.W. and D.C.; resources, G.S.; data curation, G.S.; writing—original draft preparation, H.S.; writing—review and editing, G.C.; visualization, Z.L.; supervision, G.C.; project administration, G.C.; funding acquisition, G.C. All authors have read and agreed to the published version of the manuscript.

Funding: This research received no external funding.

Institutional Review Board Statement: Not applicable.

Informed Consent Statement: Not applicable.

Data Availability Statement: All data were obtained by the author through experiment.

Conflicts of Interest: The authors declare no conflict of interest.

References

1. Hu, H.; Hong, X.; Tian, Y.; Zhang, D. Az31 magnesium alloy tube manufactured by composite forming technology including extruded-shear and bending based on finite element numerical simulation and experiments. *Int. J. Adv. Manuf. Tech.* **2021**, *115*, 2395–2402. [CrossRef]
2. He, B.; Huang, S.; He, X. Numerical simulation of gear surface hardening using the finite element method. *Int. J. Adv. Manuf. Tech.* **2014**, *74*, 665–672. [CrossRef]
3. Xin, C. Establishment and Application of Elastic-Plastic Constitutive Model for Cyclic Loading. Doctoral Dissertation, Yanshan University, Yanshan, China.
4. Li, Q.; Xin, C.; Jin, M.; Zhang, Q. Establishment and application of an anisotropic nonlinear kinematic hardening constitutive model. *Chin. J. Mech. Eng.* **2006**, *42*, 6. [CrossRef]
5. Xiao, Y.Z.; Chen, J. A review of research in macroscopic hardening models in numerical simulation of sheet metal forming. *Int. J. Plast.* **2009**, *16*, 8.
6. Ziegler, H. A modification of prager's hardening rule. *Q. Appl. Math.* **1959**, *17*, 55–65. [CrossRef]
7. Armstrong, P.J.; Frederick, C.O. A mathematical representation of the multiaxial bauschinger effect. *Mater. High Temp.* **2007**, *24*, 1–26.
8. Yu, H.Y. Comparative study on strain hardening models of thin metal sheet. *Forg. Stamp. Technol.* **2012**, *37*, 6.
9. Yu, H.Y.; Wang, Y. A combined hardening model based on chaboche theory and its application in the springback simulation. *Chin. J. Mech. Eng.* **2015**, *51*, 8. [CrossRef]
10. Han, C.; Dong, X.H. An equivalent drawbead model considering Bauschinger effect. *Forg. Stamp. Technol.* **2017**, *42*, 7.
11. Li, Q.; Jin, M.; Zou, Z.Y.; Guo, J.; Yang, C. Parameter determination and application research of mixed hardening model based on cyclic tension-compression test. *Chin. J. Mech. Eng.* **2020**, *2*, 6. [CrossRef]
12. Chen, G.; Zhao, C.C.; Yang, Z.Y.; Dong, G.J.; Cao, M.Y. Effects of Q235 Coated Tubes on Bulging Behavior of AA5052 Aluminum Alloy Base Tube with Granular Medium. *Chin. J. Mech. Eng.* **2021**, *32*, 10.
13. Boger, R.K.; Wagoner, R.H.; Barlat, F.; Lee, M.G.; Chung, K. Continuous, large strain, tension/compression testing of sheet material. *Int. J. Plast.* **2005**, *21*, 2319–2343. [CrossRef]
14. Kuwabara, T.; Kumano, Y.; Ziegelheim, J.; Kurosaki, I. Tension-compression asymmetry of phosphor bronze for electronic parts and its effect on bending behavior. *Int. J. Plast.* **2005**, *25*, 1759–1776. [CrossRef]
15. Yoshida, F.; Uemori, T. *Cyclic Plasticity Model for Accurate Simulation of Springback of Sheet Metals*; Springer Vieweg: Berlin/Heidelberg, Germany, 2015; pp. 65–66.
16. Jian, C.; Lee, W.; Hang, S.C.; Seniw, M.; Chung, K. Experimental and numerical investigation of combined isotropic-kinematic hardening behavior of sheet metals. *Int. J. Plast.* **2009**, *25*, 942–972.
17. Hang, S.C.; Lee, W.; Jian, C.; Seniw, M.; Chung, K. Experimental and Numerical Investigation of Kinematic Hardening Behavior in Sheet Metals. In Proceedings of the 10th Esaform Conference on Material Forming, Zaragoza, Spain, 18–20 April 2007; Volume 907, pp. 337–342.
18. Kurukuri, S.; Worswick, M.J.; Ghaffari Tari, D.; Mishra, R.K.; Carter, J.T. Rate sensitivity and tension-compression asymmetry in AZ31B magnesium alloy sheet. *Philos. Trans.* **2014**, *372*, 20130216. [CrossRef] [PubMed]
19. Abedini, A.; Butcher, C.; Nemcko, M.J.; Kurukuri, S.; Worswick, M.J. Constitutive characterization of a rare-earth magnesium alloy sheet (ZEK100-O) in shear loading: Studies of anisotropy and rate sensitivity. *Int. J. Mech. Sci.* **2017**, *128*, 54–69. [CrossRef]
20. Ma, M.T. A review of the Bauschinger effect in metals and alloys. *Mater. Mech. Eng.* **1986**, *10*, 15–21.
21. Wang, Y.F.; Li, C.; Ling, X.Y.; Shen, B.L.; Gao, S.J.; Ying, S.H. An overview on the Bauschinger effect in metallic materials. *China Nucl. Sci. Technol. Rep.* **2002**, *1*, 14.
22. Noma, N.; Kuwabara, T. Numerical investigation of specimen geometry for in-plane compression tests and its experimental validation. *J. Jpn. Soc. Technol. Plast.* **2012**, *53*, 574–579. [CrossRef]
23. Dong, X.H. *Principles of Metal Forming*; China Machine Press: Beijing, China, 2011; pp. 154–196.

Disclaimer/Publisher's Note: The statements, opinions and data contained in all publications are solely those of the individual author(s) and contributor(s) and not of MDPI and/or the editor(s). MDPI and/or the editor(s) disclaim responsibility for any injury to people or property resulting from any ideas, methods, instructions or products referred to in the content.

Article

An Algorithm for Real-Time Aluminum Profile Surface Defects Detection Based on Lightweight Network Structure

Junlong Tang [1,*], Shenbo Liu [1], Dongxue Zhao [1], Lijun Tang [1], Wanghui Zou [1] and Bin Zheng [2]

1. School of Physics and Electronic Science, Changsha University of Science and Technology, Changsha 410114, China
2. School of Computer and Communications Engineering, Changsha University of Science and Technology, Changsha 410114, China
* Correspondence: tangjl@csust.edu.cn

Abstract: Surface defects, which often occur during the production of aluminum profiles, can directly affect the quality of aluminum profiles, and should be monitored in real time. This paper proposes an effective, lightweight detection method for aluminum profiles to realize real-time surface defect detection with ensured detection accuracy. Based on the YOLOv5s framework, a lightweight network model is designed by adding the attention mechanism and depth-separable convolution for the detection of aluminum. The lightweight network model improves the limitations of the YOLOv5s framework regarding to its detection accuracy and detection speed. The backbone network GCANet is built based on the Ghost module, in which the Attention mechanism module is embedded in the AC3Ghost module. A compression of the backbone network is achieved, and more channel information is focused on. The model size is further reduced by compressing the Neck network using a deep separable convolution. The experimental results show that, compared to YOLOv5s, the proposed method improves the mAP by 1.76%, reduces the model size by 52.08%, and increases the detection speed by a factor of two. Furthermore, the detection speed can reach 17.4 FPS on Nvidia Jeston Nano's edge test, which achieves real-time detection. It also provides the possibility of embedding devices for real-time industrial inspection.

Keywords: real-time detection; lightweight network structure; YOLOv5s; attention mechanism; edge computing

1. Introduction

Due to their excellent thermal conductivity and moisture resistance, aluminum profiles have become an important primary material for buildings, vehicles, ships, houses, and other fields. With the rapid development of related industries, the demand for high-quality aluminum profiles is also increasing. Surface defects on aluminum profiles directly affect the quality of products. Therefore, it is significant to detect those defects during their production.

It is difficult to use traditional manual visual inspection to ensure the accuracy of inspection results and inspection efficiency because manual processes can produce a series of problems, such as inefficiency and human physiological fatigue [1]. Some scholars have applied machine learning methods for industrial defect recognition to solve those problems. Yu et al. [2] utilized SVM (support vector machine) to classify wood surface defects. The recognition accuracy of the back propagation neural network model proposed by the authors was 92.7% and 92.0% in the training and test sets, respectively. Hu et al. [3] proposed an algorithm based on ellipse fitting with distance thresholding to detect crater defects on steel shell surface. Elliptical fitting of the extracted inner circle curve was performed, and thus there was high accuracy and detection efficiency for crater defects. You et al. [4] identified crack defects of 0.15 mm using the C-scanning method. This approach can only identify crack defects and has its limitations. K et al. [5] realized the detection of internal defects in carbon-fiber-reinforced plastics and glass-fiber-reinforced plastics using recurrence methods and C-scans. Chen et al. [6] presented smooth filtering to detect steel plate surface defects. Wang et al. [7] use SUSAN operator to detect the edges of the foil image and obtain the threshold aluminum foil image to determine the effective area of the foil in the image. The localization and identification of defects on the surface of aluminum foil were achieved. Although the abovementioned works have achieved some good results in surface defect detection, there are still some limitations, such as poor robustness and weak adaptability.

With the convolutional neural networks (CNNs) proposed, deep learning, which overcomes the limitations of machine learning methods, has been widely used for surface defect detection [8]. Deep-learning-based target detection algorithms are mainly divided into two categories. One is a two-stage classification, and the representative algorithms include R-CNN (regions with CNN features) [9], Fast R-CNN(fast region-based CNN) [10], and Faster R-CNN [11]. These algorithms are applied to the creation and the classification of candidate boxes. The other is single-stage classification, and the representative algorithms include SSD (Single Shot MultiBox Detector) [12], YOLO (You Only Look Once) [13–16], CenterNet [17], and Retinanet [18]. These algorithms generate class probability values and coordinates of the position of the target object during the creation of a candidate frame. The final detection result can be obtained directly after detecting the target. Fu et al. [19] proposed an end-to-end model based on SqueezeNet to achieve steel strip detection under inhomogeneous illumination with a detection speed of more than 100 fps. However, a dataset with insignificant difference in defect target size was used. Li et al. [20] have improved the network structure of YOLO and achieved an accurate detection of steel strips with 95.86% mAP for defects. Yang et al. [21] have realized the detection of surface defects in automotive pipe joints based on wavelet decomposition and convolutional neural networks. Amin et al. [22] fulfilled the detection of surface defects in steel based on U-NET [23]. Defects can be detected quickly, but the detection accuracy is only 0.731. Zhang et al. [24] proposed the MRSDI-CNN algorithm, which combined SSD with YOLOv3 for the recognition of surface defects on steel rails. The detection speed was improved to a certain extent, but the real-time detection is not realized on embedded devices. Chen et al. [25] have fulfilled the recognition of steel rail surface defects based on Faster R-CNN with 97.8% mAP of blemishes. However, it is not real-time to detection of surface defects. Y et al. [26] implemented the defect depth detection of 3D woven composites using Fully Convolutional Neural Network and recurrence methods. Zhou et al. [27] implemented microtubule defect detection on wafer surface by embedding DA attention module in YOLOv5. The algorithm model has good detection capability on small target defects; however, it is large in size and cannot achieve real-time detection. A CNN-based detection algorithm can be accurate regarding aluminum surface defects, but the detection speed cannot meet industrial inspection needs.

With the development of lightweight network technology, many scholars have realized the real-time detection of surface defects of aluminum profiles by adding a lightweight network to the detection algorithm. Ma et al. [28] implemented the detection of surface defects on aluminum strips by embedding a Ghost module with union attention mechanism in YOLOv4 network. The method achieved an mAP value of 94.68%, a model volume reduction of 80.41%, a threefold increase in detection speed, and better performance than the YOLOv4 model. However, the model size of the algorithm is 48.5 MB and the detection speed is 20.749 f/s, so there is still room for improvement. Wang et al. [29] proposed an improved MS-YOLOv5 model based on the YOLOv5 algorithm. A multi-stream network was present as the first detection head of the algorithm, and the Neck layer was optimized. The model recognition ability and model localization extraction at different sizes were improved. However, the mAP of the model is 87.4%, which cannot meet the needs of industrial inspection. Yang et al. [30] achieved accurate identification of dirty spots by embedding FPN structures in Faster R-CNN to improve the model's ability to extract feature information about defects. There were strong limitations for the algorithm. It was used to only identify defects within a single category under specific conditions, and is not suitable for surface defect detection of aluminum profiles in industrial manufacturing processes. Li et al. [31] implemented the detection of 10 kinds of aluminum profile surface defects based on migration learning, and the classification accuracy reached 98.47%. However, the speed of detection was not mentioned. Wu et al. [32] proposed a defect detection model based on YOLO X, which replaced the original CSP-DarkNet with CSP-ResNeXt and integrated an attention mechanism. The algorithm achieved 90.69% mAP with a detection speed of 33.6 FPS. Although the above work has achieved some good results in the detection of aluminum surface defects, the detection speed and detection accuracy still need to be further improved.

Deep-learning-based defect detection algorithms can achieve high accuracy rates for specific datasets. As the size of defects on the surface of aluminum profiles varies greatly, those methods fail to achieve good results, and real-time detection is difficult to implement in embedded devices. Therefore, this study designs a novel lightweight network based on the YOLOv5s algorithm to realize a real-time defect detection of aluminum profile surfaces on an embedded system with promising detection accuracy. Depth-separable convolution and Ghostconv are used to compress the lightweight network model, and an attention mechanism is embedded in the backbone network to increase the attention of the network to channel information. As a result, the detection accuracy of the algorithm is improved. The proposed algorithm of the novel lightweight network can reach an accurate real-time detection on embedded devices in this study. Specific innovation points are as follows:

(1) A lightweight model based on YOLOv5s is proposed. Compared with the YOLOv5s algorithm, the proposed model greatly improves the speed and accuracy, while the model size is greatly reduced to facilitate the deployment of edge devices.

(2) GCANet is constructed by combining a Ghost module and attention mechanism, which significantly improves the model detection speed, reduces memory consumption, and ensures model accuracy. Moreover, the Attention mechanism module is embedded in the backbone network, which mainly enhances the ability of the backbone network to focus on the channel information.

(3) In the neck network of the lightweight model, the regular convolution is replaced by the depthwise separable convolution, which greatly compresses the model size and further improves the detection speed.

2. Image Preprocessing and Datasets

Insufficient training samples during the training process can lead to low detection accuracy, overfitting, and low robustness. The number of images is increased by appropriate enhancement of the original images to effectively solve the problem of insufficient training samples [33]. Four typical types of defects on the surface of an aluminum profile are scuffing, soiling, folds, and pinholes, as shown in Figure 1. For the dataset, the pixel size of each image is 640 × 480. A pinhole is caused by the formation of tiny pores during the solidification of aluminum, with an average pixel size of 20 × 20. According to the definition in the literature [34], a pinhole with less than 1.23% of annotated pixels is a small object. Dirt is introduced by the contamination of equipment lubricants. Scratches are caused by relative friction between aluminum and equipment during processing and production. Folds are caused by unbalanced forces during aluminum processing and production. Lin et al. [35] have improved the robustness of defect detection using Gaussian filtering for noise reduction regarding the target. Simonyan et al. [36] processed the images by random flip, rotate, and crop to effectively expand the dataset. This easily leads to missed detection and false detection for low-resolution small targets, the presence of few available features, and high positioning accuracy requirements and aggregation. To enhance the semantic information of small targets, this paper adopts noise reduction by utilizing Gaussian low-pass filtering during copy-pasting in specific regions. The pinholes generated by copy-pasting and Gaussian blur techniques have more effective feature information.

Following this, the images are expanded by random flip, rotation, and cutout. The cutout is able to further enhance the localization capability of the model by requiring the model to identify objects from a local view and adding information about other samples to the cut region. Color space transformation generally eliminates lighting, luminance and color differences. These image preprocessing operations increase the number of training datasets to make them as diverse as possible, which in turn improves the generalization ability and robustness of the model. The results of the expanded images are shown in Figure 2. After image enhancement and expansion, the dataset reached 4400 images. Among the dataset, 3600 images are randomly selected for the training set, 400 images for the testing set, and the remaining 400 images for the validation set.

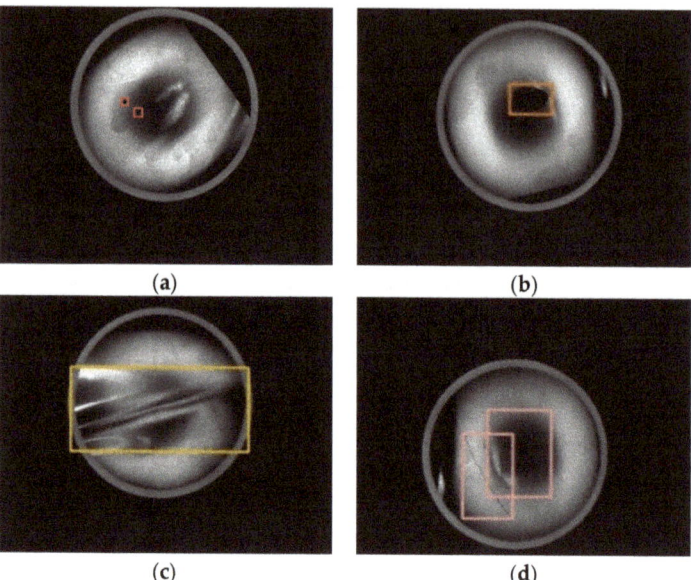

Figure 1. The four defects of aluminum profiles: (**a**) pinhole; (**b**) dirt; (**c**) fold; (**d**) scratch.

Figure 2. *Cont.*

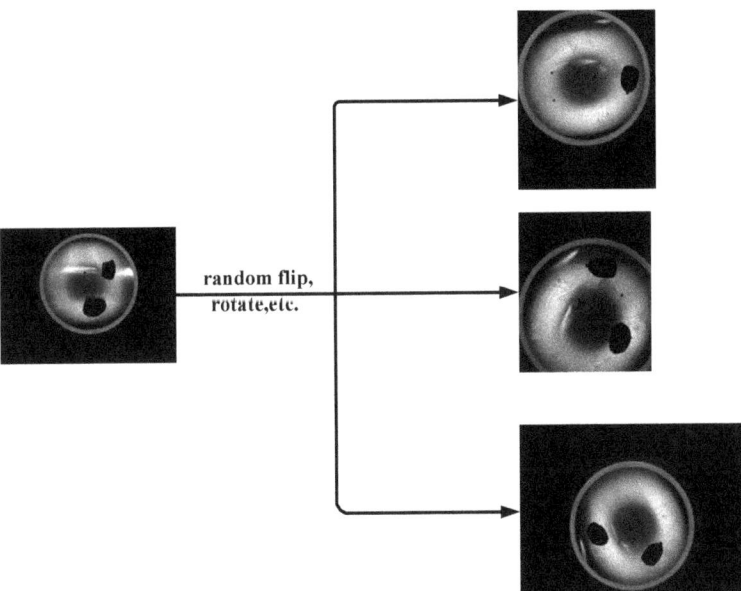

Figure 2. Images were obtained using the expansion technique.

3. Description of Methodology

3.1. Network Architecture

In this study, a lightweight model network structure based on the four basic structural frameworks of YOLOv5s is proposed, and consists of Input, GCANet backbone, Neck, and Prediction. Through the lightweight modules with embedded attention mechanisms, real-time accurate detection of surface defects on aluminum profiles is achieved.

Figure 3 shows the lightweight model network structure based on YOLOv5s. In the Input layer, the input image is resized to 640 × 640 × 3, and is input to the GCANet backbone. The attention mechanism is embedded in the C3Ghost module to improve ability of the model to focus on channel information and spatial information. Three scale feature maps, (80 × 80) (40 × 40) (20 × 20), are extracted at different levels. Following this, based on the DwConv module, the images are inputted to Neck for further compression of the model. Finally, detection is performed in Prediction.

3.2. GCANet Backbone Structure

The GCANet backbone architecture consists of the CBL module, Ghost module, and AC3Ghost module. The CBL module consists of Conv, BatchNorm, and Leaky relu. The ghost module is from GhostNet, proposed by Huawei in 2020. Compared to traditional convolution, Ghost convolution is divided into two steps, which can effectively reduce the amount of computation and number of parameters. Firstly, the standard convolution is used to compute and obtain m feature maps with fewer channel features, then s feature maps are generated using cheap linear operations. Secondly, the two feature maps are concatenated to obtain the new output of $m \cdot s$ feature maps. The structure of the Ghost module is shown in Figure 4. In standard convolution, the number of convolution kernels is assumed to be n, the size of the input feature map is $h \cdot w \cdot c$, the output feature map is $n \cdot h' \cdot w'$, and the convolution kernel is $k \cdot k$. The model floating-point computations for standard convolution and Ghost convolution are q_1 and q_2, respectively.

$$q_1 = n \cdot h' \cdot w' \cdot c \cdot k \cdot k \tag{1}$$

$$q_2 = \frac{n}{s} \cdot h' \cdot w' \cdot c \cdot k \cdot k + (s-1) \cdot \frac{n}{s} \cdot h' \cdot w' \cdot d \cdot d \tag{2}$$

where c denotes the number of channels of the input image, $k \cdot k$ denotes the size of the convolution kernel of the standard convolution operation, h and w are the height and width of the original feature map by Ghost convolution, h' and w' denotes the height and width of the original feature map generated by Ghost convolution, $d \cdot d$ is the size of the convolution kernel of the linear operation, and $s << c$.

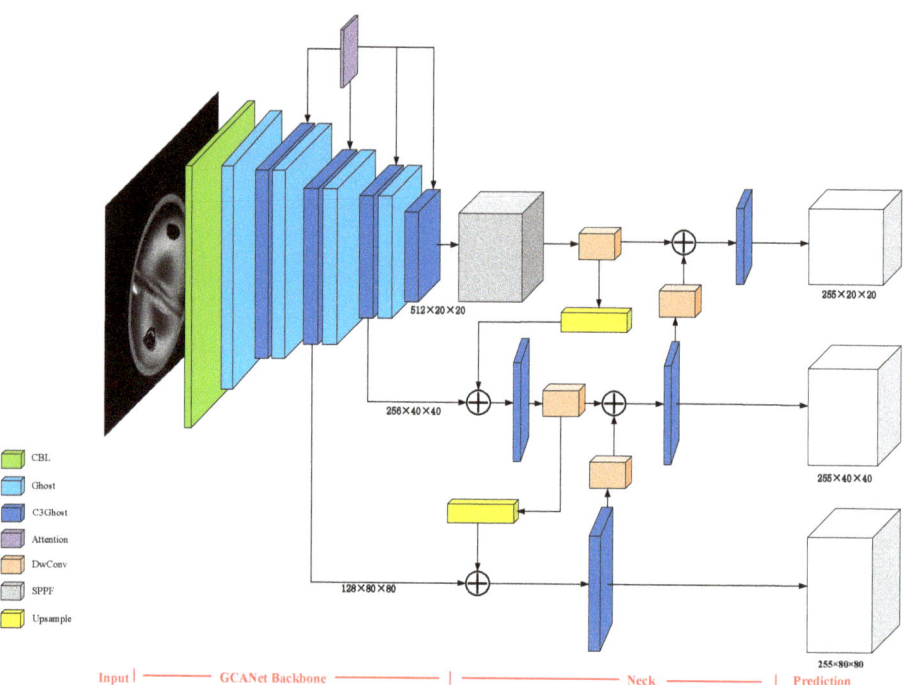

Figure 3. The lightweight model network structure based on YOLOv5s.

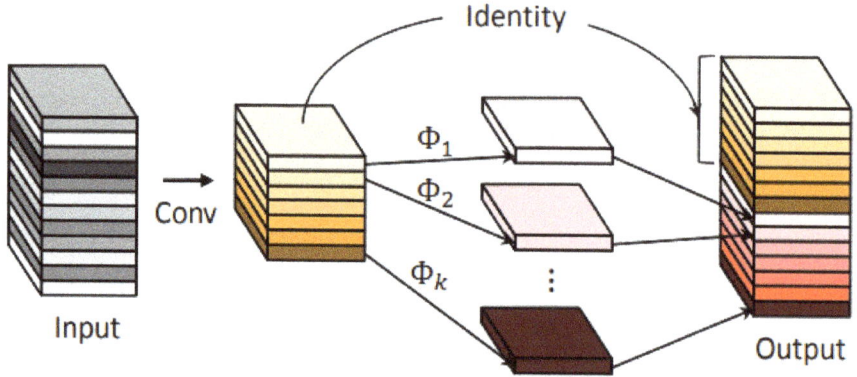

Figure 4. Ghost module structure.

The comparison of the computation of the standard convolution operation and the Ghost module is shown in (3). A comparison of the parametric quantities of the two convolutions is shown in (4). From Equations (3) and (4), it can be seen that when k and d are equal in size, the number of parameters and the computational effort for feature extraction of Ghost convolution is about $1/s$ for that of the standard convolution.

$$r_s = \frac{n \cdot h' \cdot w' \cdot c \cdot k \cdot k}{\frac{n}{s} \cdot h' \cdot w' \cdot c \cdot k \cdot k + (s-1) \cdot \frac{n}{s} \cdot h' \cdot w' \cdot d \cdot d} \\ = \frac{c \cdot k \cdot k}{\frac{1}{s} \cdot c \cdot k \cdot k + \frac{s-1}{s} \cdot d \cdot d} \approx \frac{s \cdot c}{s + c - 1} \approx s \quad (3)$$

$$r_c = \frac{n \cdot c \cdot k \cdot k}{\frac{n}{s} \cdot c \cdot k \cdot k + (s-1) \cdot \frac{n}{s} \cdot d \cdot d} \approx \frac{s \cdot c}{s + c - 1} \approx s \quad (4)$$

3.3. AC3Ghost Structure

Adding attention mechanisms to neural networks can effectively improve the performance of network feature extraction. Hu et al. [37] proposed an SE attention mechanism to establish spatial correlation in feature maps. Hou et al. [38] proposed the CA attention mechanism to integrate spatial coordinate information into feature maps effectively. Woo [39] proposed the CBAM attention mechanism to pay attention to channel and spatial information. To effectively utilize the channel and spatial information, this paper proposes an AC3Ghost module consisting of the CBAM attention mechanism and C3Ghost module, and the CBAM attention mechanism is embedded in C3Ghost. The structure of the AC3Ghost module is shown in Figure 5. When the data processed by the Ghost are input to AC3Ghost, the AC3Ghost module is divided into two branches to process in parallel, one for hierarchical feature fusion by multiple Ghost Bottleneck stacks and three 1 × 1 convolution modules, and the other for reducing the number of channels by only one 1 × 1 convolution module. Following this, feature maps of the two branches are fused as output feature maps by concat, and the CBAM attention mechanism focuses on the channel and spatial information. Finally, it passes through a 1 × 1 convolution module.

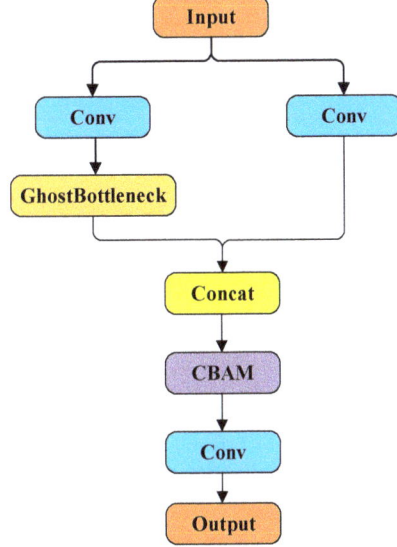

Figure 5. AC3Ghost module Structure.

3.4. DwConv Module

Depthwise separable convolution (DwConv) was proposed in MobileNet [40] in 2017. DwConv reduces the number of parameters needed during the convolution calculation and improves the efficiency of convolution by splitting the standard convolution in the spatial dimension and channel dimension. As shown in Figure 6, the DwConv module structure is divided into two main processes of Depthwise Convolution and Pointwise Convolution. One convolution kernel of Depthwise Convolution is responsible for one channel. One channel is convolved by only one convolution kernel. The process producing the Pointwise Convolution is very similar to regular convolution. It has a convolution kernel size of 1 × 1, and is weighted in the direction of the map depth from the previous step to generate a new feature map. The computational complexity of a regular convolution C_{Conv} is shown in Equation (5), and the computational complexity of a depth-separable convolution $C_{separableConv}$ is shown in Equation (6). The ratio of the computational cost of deep separable convolution to that of standard convolution is shown in Equation (7). Experiments show [32] that the computation is 8–9 times less than the standard convolution when the convolution kernel size of DwConv is set to 3 × 3.

$$C_{Conv} = D_{out\,1} \cdot D_{out\,2} \cdot D_{k1} \cdot D_{k2} \cdot C_{out} \cdot C_{in} \tag{5}$$

$$C_{separableConv} = D_{out\,1} \cdot D_{out\,2} \cdot D_{k1} \cdot D_{k2} \cdot C_{in} + D_{out\,1} \cdot D_{out\,2} \cdot C_{out} \cdot C_{in} \tag{6}$$

$$\frac{C_{Separable\,Conv}}{C_{Conv}} = \frac{D_{out\,1} \cdot D_{out\,2} \cdot D_{k1} \cdot D_{k2} \cdot C_{in} + D_{out\,1} \cdot D_{outt} \cdot C_{out} \cdot C_{in}}{D_{outl} \cdot D_{out\,2} \cdot D_{k1} \cdot D_{k2} \cdot C_{out} \cdot C_{in}} \tag{7}$$

where D_{in1}, D_{in2} are the input dimensions, $D_{out\,1}$, $D_{out\,2}$ are the output dimensions, D_{k1}, D_{k2} are the convolution kernel size, C_{in} is the number of input channels, and C_{out} is the number of output channels.

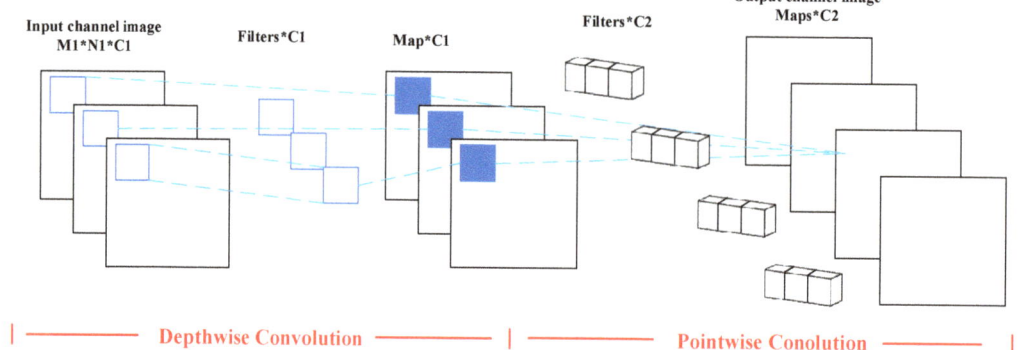

Figure 6. DwConv module Structure.

4. Experiments and Discussion

4.1. Experimental Environment

All experiments were performed on a CPU with NVIDIA GeForce RTX 3090 24 GB GPU and Intel i7-12700. The computing software environment was set to python 3.8, CUDA Version 11.6, and the compiler was PyTorch 1.11. In the network training, this study took a batch size of 64, a learning rate of 0.001, an epoch of 500, and an SGD momentum is 0.937.

4.2. Evaluation of Model Performance

Average precision (AP) indicates the accuracy of categories. The mAP is the average of AP, representing the averaged accuracy of all categories. mAP@0.5 indicates the mAP with an IOU greater than 0.5, while mAP@0.5:0.95 indicates the mAP with an IOU at [0.5,0.95]. The IOU is shown in Equation (8). Precision measures the exactness of classification, as shown in Equation (9). AP and mAP are shown in Equations (10) and (11), respectively. In Equations (8)–(11), box_{gt} is the ground truth of the defect; box_p is the predicted area of the defect; TP and FP are, respectively, the numbers of true-positive cases and false-positive cases; and n is the number of detection classes. TN and FN are, respectively, the numbers of true-negative cases and false-negative cases, and FPS is used to evaluate the detection speed of the model. In this paper, mAP@0.5, mAP@0.5:0.95, FPS, and model size are chosen to evaluate the experimental method.

$$IOU(box_{gt}, box_p) = \frac{|box_{gt} \cap box_p|}{|box_{gt} \cup box_p|} \tag{8}$$

$$Precision = \frac{TP}{(TP+FP)} \tag{9}$$

$$AP = \frac{\sum_{i=1}^{n} P_i}{n} \tag{10}$$

$$mAP = \frac{\sum_{i=1}^{k} AP_i}{k} \tag{11}$$

4.3. Test Result of Defect Detection

This paper adopts a fivefold cross-validation method to test the algorithm's accuracy. The dataset is divided into five groups, four of them are used as training data, and the remaining one as the test data. Experiments were conducted, with the results shown in Table 1. The experiments with the best mAP in Table 1 will also be used for the subsequent comparison experiments. The experiments show that the Precision, Recall and AP@.5 of the fold reach 99.45, 100, and 99.37, respectively, which has a better performance, because the folds are distinctive, fixed in shape, and easy to locate. Similarly, the accuracy and detection rate of this paper's method for both types of scratches and dirt are very high, and there are few cases of missed detection. These two defects have distinctive features and slight shape variations, and are less subject to interference from the background and other defects. However, for Pinhole, which has fewer pixel points, the algorithm captures and displays less texture information. In this paper, copy-pasting and Gaussian blur techniques are used to generate more feature information to meet the model training requirements. The results showed that the AP of the pinhole reached 82.45%. The various defect detection visual results of image detection are shown in Figure 7. All four defects can be detected accurately with a confidence score above 0.8.

Table 1. Test results for four types of defect detection.

	Pinhole	Scratch	Dirt	Fold
Precision	83.24	98.86	99.11	99.45
Recall	77.21	99.39	99.22	100.00
AP@0.5:0.95	51.94	79.80	81.10	80.60
AP@0.5	82.45	98.77	98.80	99.37
mAP@0.5:0.95			73.36	
mAP@0.5			94.85	

4.4. Comparison of the Effect of Different Defect Detection Algorithms

The proposed method in this paper is compared with the following single-stage algorithms: SSD, YOLOv3-tiny, YOLOv4-tiny, YOLOv5-Mobilenetv3, YOLOv5-Shufflenetv2, and YOLOv5. The visualization results of the detection results of different algorithms are shown in Figure 8. The features of mAP@0.5, mAP@0.5:0.95, Model Size, and Detection Speed are compared, as shown in Table 2. Compared with the other six algorithms, the proposed algorithm has the best performance in mAP@0.5 and mAP@0.5:0.95 with 94.85 and 73.36. YOLOv5-Mobilenetv3 and YOLOv5-Shufflenetv2 have good detection speed, but the mAP@0.5 is below 90% and cannot achieve the accurate detection of surface defects in aluminum profiles. From the analysis of the results, it can be seen that the proposed method possesses great advantages in the accuracy and real-time detection of surface defects in aluminum profiles, and the detection speed reaches 136 FPS.

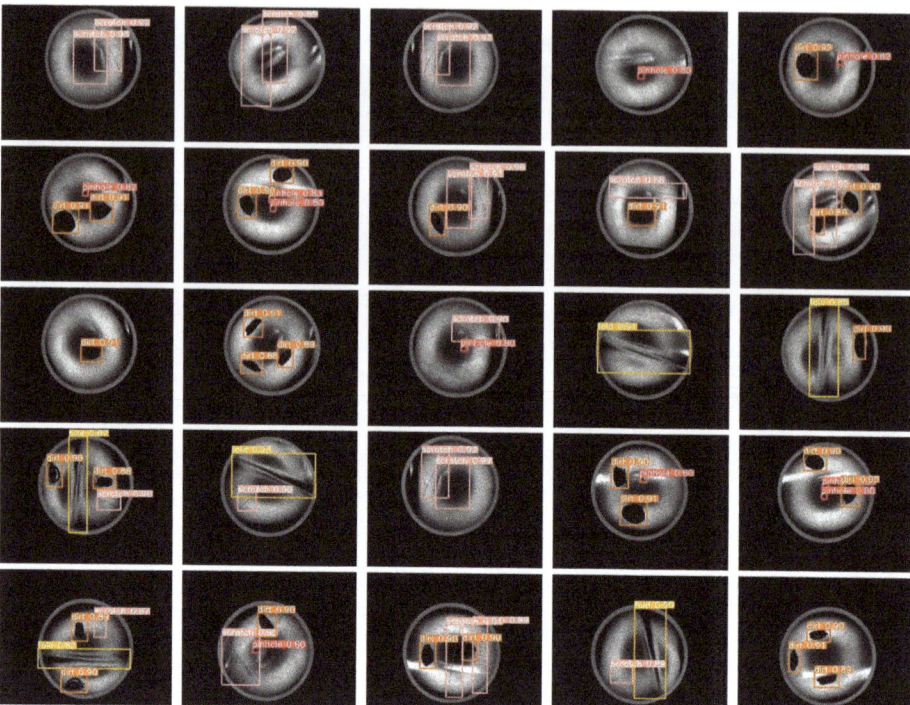

Figure 7. Visual inspection results for four types of defects.

Table 2. Results of different object detection models.

Algorithm	mAP@0.5%	mAP@0.5:0.95%	Model Size (MB)	Detection Speed (FPS)
SSD	67.86	42.76	91.09	39
YOLOv3-tiny	85.82	65.89	33.19	66
YOLOv4-tiny	89.31	67.86	23.09	86
YOLOv5s	93.09	72.58	14.40	75
YOLOv5-Mobilenetv3	89.71	67.78	5.96	158
YOLOv5-Shufflenetv2	89.65	66.86	3.30	176
Ours	94.85	73.36	7.80	136

4.5. Ablation Study

To understand well the contribution of the improved module to the defect detection effect, a large number of ablation experiments were performed to further verify the YOLOv5s. The results of the ablation study are exhibited in Table 3, where the baseline is the YOLOv5s after data enhancement is performed.

Table 3. Effects of various design modules on surface image.

Description	Model Size (MB)	mAP@0.5 (%)	Change (%)
YOLOv5s	14.40	90.32	–
Baseline	14.40	93.09	+2.77
+GCANet Backbone Network Structure	7.90	93.90	+0.81
+AC3Ghost Structural	8.20	94.96	+1.06
+DwConv Module	7.80	94.84	−0.12

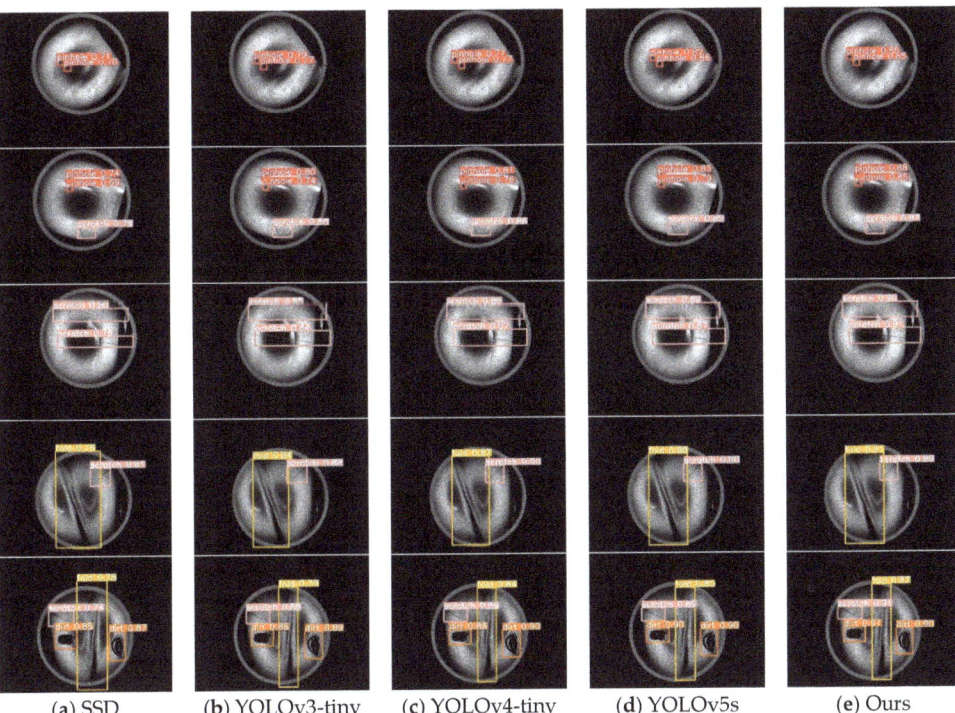

Figure 8. Visual comparison of the results of the five algorithms on the dataset.

As shown in Table 3, image preprocessing has a good effect on the model's accuracy. Image preprocessing can increase the number and diversity of images and enrich the characteristics of defects. The mAP@0.5 has been increased by 2.77%. Adding a GCANet Backbone Network Structure on top of the proposed lightweight model can significantly reduce the model's size and improve the model's detection capability, and mAP is increased by 0.81%, because ghostconv in GCANet network can reduce a lot of redundant information by linear transformation. Continually adding the AC3Ghost structure after the GCANet Backbone, the detection accuracy of the model is improved further, and the model mAP is increased by 1.06%, but the model size is also slightly increased by 0.3 MB. The DwConv module improves the standard convolution by splitting the correlation between spatial and channel dimensions, reducing the number of parameters needed for the convolution calculation and improving the efficiency of the convolution. For DwConv Module, the model size is reduced by 0.4 MB, and the mAP was slightly dropped from 94.96% to 94.84%, only decreased by 0.12%. At the end, the overall mAP of the model was increased by 0.94%. These results verify that the proposed network model is effective for detecting surface defects in aluminum strips, and that the detection accuracy can be ensured while the model's size is significantly reduced.

4.6. Edge Testing

To verify the detection of the improved algorithm on the embedded platform, an edge surface defect-detection system was built, as shown in Figure 9 and Table 4. The system consists of an LED light source, a CCD image sensor, a 7-inch touch screen, an Nvidia Jeston Nano, an encoder, a conveyor belt, and two power supplies. The testing sample arrives at the center of the CCD image sensor via a conveyor belt. The CCD image sensor detects the sample and displays the results on the Touch screen. The experimental results show that the edge surface defect-detection system can achieve real-time detection of surface defects on aluminum profiles with a detection speed of 17.4 FPS with good robustness. In summary, the proposed method takes into account both accuracy and real-time operation, and can achieve good detection results in the detection of defects in the industrial production of aluminum profiles.

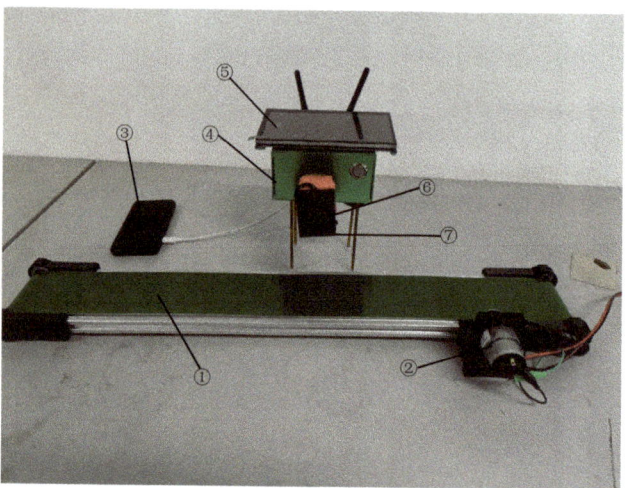

Figure 9. Edge surface defect detection system.

Table 4. Serial number and name of component in detection system.

Number	Component Name
1	Conveyor belt
2	Encoder
3	Power supply
4	Nvidia Jeston Nano
5	Touch screen
6	CCD
7	LED light source

5. Conclusions

Surface defects on aluminum profiles directly affect their quality. Advanced inspection processes and methods can ensure the accuracy of detection results with high-efficiency detection process. In this paper, a lightweight network model was proposed by adding an attention mechanism and depth-separable convolution for the detection of surface defects in an aluminum strip. By combining the ghost module and the attention mechanism, a new backbone was built. The model size was reduced by 6.2 MB and the mAP was increased by 4.64%. In the neck network of the lightweight model, the regular convolution was replaced by the deeply separable convolution. The model size was further compressed to 7.8 MB. Compared with other object detections and lightweight model experiments, the proposed algorithm has better real-time performance and accuracy than other single-stage detection algorithms. The detection accuracy mAP@0.5 is 94.85, mAP@0.5:0.95 is 73.36, and the detection speed is 136.98 FPS. Furthermore, in edge testing, the proposed algorithm in the present work can achieve real-time detection. Experiments show that it can detect the surface defects of aluminum profiles in real time with guaranteed high accuracy. In addition, the algorithm has high scalability and can be extended to other fields such as PCB surface defects. In the future, this research work will be focused on continuing to improve pinhole detection accuracy in complex contexts. At the same time, we will continue to carry out research efforts on the types of aluminum profile surface defects to achieve the identification of more defect types.

Author Contributions: Conceptualization, J.T. and S.L.; methodology, J.T.; software, J.T.; validation, J.T., S.L. and D.Z.; formal analysis, J.T.; investigation, J.T.; resources, W.Z.; data curation, J.T.; writing—original draft preparation, B.Z.; writing—review and editing, D.Z.; visualization, J.T.; supervision, L.T.; project administration, J.T.; funding acquisition, J.T. All authors have read and agreed to the published version of the manuscript.

Funding: This research was funded by the Open Research Fund of Hunan Provincial Key Laboratory of Flexible Electronic Materials Genome Engineering under grant (No. 202015).

Institutional Review Board Statement: Not applicable.

Informed Consent Statement: Not applicable.

Data Availability Statement: The data presented in this study are available on request from the corresponding author.

Acknowledgments: The authors would like to acknowledge the anonymous reviewers and editors whose thoughtful comments helped to improve this manuscript.

Conflicts of Interest: The authors declare they have no conflict of interest.

References

1. Thomas, S.S.; Gupta, S.; Subramanian, V.K. Smart surveillance based on video summarization. In Proceedings of the IEEE Region 10 Symposium (TENSYMP), Cochin, India, 14–16 July 2017; pp. 1–5.
2. Yu, H.; Liang, Y.; Liang, H.; Zhang, Y. Recognition of wood surface defects with near infrared spectroscopy and machine vision. *J. For. Res.* **2019**, *30*, 2379–2386. [CrossRef]
3. Hu, H.; Xu, D.; Zheng, X.; Zhang, B. Pit defect detection on steel shell end face based on machine vision. In Proceedings of the 2020 IEEE 4th Information Technology, Networking, Electronic and Automation Control Conference (ITNEC), Chongqing, China, 12–14 June 2020.
4. You, Y.; Xiao, Z.; Chun-Guang, X.U.; Xiao, D.G.; Yi-Xin, H.U.; Pei-Lu, L.I.; Guo, C.Z. Robotic NDT for Turbine Blades Based on the Transverse Waves. In Proceedings of the 2017 Far East NDT New Technology & Application Forum (FENDT), Xi'an, China, 22–24 June 2017; pp. 18–22.
5. Ciecielag, K.; Kecik, K.; Skoczylas, A.; Matuszak, J.; Korzec, I.; Zaleski, R. Non-Destructive Detection of Real Defects in Polymer Composites by Ultrasonic Testing and Recurrence Analysis. *Materials* **2022**, *15*, 7335. [CrossRef] [PubMed]
6. Chen, N.; Sun, J.; Wang, X.; Huang, Y.; Li, Y.; Guo, C. Research on surface defect detection and grinding path planning of steel plate based on machine vision. In Proceedings of the 2019 14th IEEE Conference on Industrial Electronics and Applications (ICIEA), Xi'an, China, 19–21 June 2019; pp. 1748–1753.
7. Hui, W.; Chunhua, G.; Xiangxu, X.; Yongfa, L.; Yuji, W. Study on Edge Detection Method of Aluminum Foil Image. In Proceedings of the 2017 International Conference on Computer Systems, Electronics and Control (ICCSEC), Dalian, China, 25–27 December 2017; pp. 1008–1010.
8. Indolia, S.; Goswami, A.; Mishra, S.; Asopa, P. Conceptual understanding of convolutional neural network-a deep learning approach. *Proc. Comput. Sci.* **2018**, *132*, 679–688. [CrossRef]
9. Girshick, R.; Donahue, J.; Darrell, T.; Malik, J. Rich feature hierarchies for accurate object detection and semantic segmentation. In Proceedings of the IEEE Computer Society Conference on Computer Vision and Pattern Recognition, Columbus, OH, USA, 23–28 June 2014; pp. 580–587.
10. Girshick, R. Fast R-CNN. In Proceedings of the IEEE International Conference on Computer Vision, Washington, DC, USA, 7–13 December 2015; pp. 1440–1448.
11. Ren, S.; He, K.; Girshick, R.; Sun, J. Faster R-CNN: Towards real-time object detection with region proposal networks. *IEEE Trans. Pattern Anal. Mach. Intell.* **2017**, *39*, 1137–1149. [CrossRef] [PubMed]
12. Liu, W.; Anguelov, D.; Erhan, D.; Szegedy, C.; Reed, S.; Fu, C.; Berg, A.C. SSD: Single shot multibox detector. In Proceedings of the Computer Vision–ECCV 2016: 14th European Conference, Amsterdam, The Netherlands, 11–14 October 2016; pp. 21–37.
13. Redmon, J.; Divvala, S.; Girshick, R.; Farhadi, A. You only look once: Unified real-time object detection. In Proceedings of the IEEE Conference on Computer Vision and Pattern Recognition, Las Vegas, NV, USA, 26 June–1 July 2016; pp. 779–788.
14. Redmon, J.; Farhadi, A. YOLO9000: Better faster stronger. In Proceedings of the IEEE Conference on Computer Vision and Pattern Recognition, San Juan, PR, USA, 17–19 June 2017; pp. 7263–7271.
15. Redmon, J.; Farhadi, A. YOLOv3: An incremental improvement. *arXiv* **2018**, arXiv:1804.02767.
16. Bochkovskiy, A.; Wang, C.-Y.; Liao, H.-Y.M. YOLOv4: Optimal speed and accuracy of object detection. *arXiv* **2020**, arXiv:2004.10934.

17. Duan, K.; Bai, S.; Xie, L.; Qi, H.; Huang, Q.; Tian, Q. CenterNet: Keypoint triplets for object detection. In Proceedings of the IEEE/CVF International Conference on Computer Vision, Seoul, Republic of Korea, 28 October 2019; pp. 6569–6578.
18. Lin, T.Y.; Goyal, P.; Girshick, R.; He, K.; Dollár, P. Focal loss fordense object detection. In Proceedings of the IEEE International Conference on Computer Vision, Venice, Italy, 22–29 October 2017; pp. 2980–2988.
19. Cheng, X.; Yu, J. RetinaNet with difference channel attention and adaptively spatial feature fusion for steel surface defect detection. *IEEE Trans. Instrum. Meas.* **2021**, *70*, 2503911. [CrossRef]
20. Fu, G.; Sun, P.; Zhu, W.; Yang, J.; Cao, Y.; Yang, M.Y.; Cao, Y. A deep-learning-based approach for fast and robust steel surface defects classification. *Opt. Lasers Eng.* **2019**, *121*, 397–405. [CrossRef]
21. Li, J.; Su, Z.; Geng, J.; Yin, Y. Real-time detection of steel strip surface defects based on improved YOLO detection network. *IFAC-PapersOnLine* **2018**, *51*, 76–81. [CrossRef]
22. Yang, Z.; Zhang, M.; Li, C.; Meng, Z.; Li, Y.; Chen, Y.; Liu, L. Image Classification for Automobile Pipe Joints Surface Defect Detection Using Wavelet Decomposition and Convolutional Neural Network. *IEEE Access* **2022**, *10*, 77191–77204. [CrossRef]
23. Amin, D.; Akhter, S. Deep Learning-Based Defect Detection System in Steel Sheet Surfaces. In *2020 IEEE Region 10 Symposium (TENSYMP)*; IEEE: Piscataway, NJ, USA, 2020; pp. 444–448.
24. Ronneberger, O.; Fischer, P.; Brox, T. U-NET: Convolutional networks for biomedical image segmentation. In Proceedings of the International Conference on Medical Image Computing and Computer-Assisted Intervention, Munich, Germany, 5–9 October 2015; Volume 2, pp. 234–241.
25. Zhang, H.; Song, Y.; Chen, Y.; Zhong, H.; Liu, L.; Wang, Y.; Wu, Q.J. MRSDI-CNN: Multi-Model Rail Surface Defect Inspection System Based on Convolutional Neural Networks. In *IEEE Transactions on Intelligent Transportation Systems*; IEEE: Piscataway, NJ, USA, 2022; Volume 23, pp. 11162–11177.
26. Chen, X.; Zhang, H. Rail Surface Defects Detection Based on Faster R-CNN. In Proceedings of the 2020 International Conference on Artificial Intelligence and Electromechanical Automation (AIEA), Tianjin, China, 26–28 June 2020; pp. 819–822.
27. Guo, Y.; Xiao, Z.; Geng, L.; Wu, J.; Zhang, F.; Liu, Y.; Wang, W. Fully Convolutional Neural Network With GRU for 3D Braided Composite Material Flaw Detection. *IEEE Access* **2019**, *7*, 151180–151188. [CrossRef]
28. Zhou, N.; Liu, Z.; Zhou, J. Yolov5-based defect detection for wafer surface micropipe. In Proceedings of the 3rd International Conference on Information Science, Parallel and Distributed Systems, ISPDS, Guangzhou, China, 22–24 July 2022; pp. 165–169.
29. Ma, Z.; Li, Y.; Huang, M.; Huang, Q.; Cheng, J.; Tang, S. Automated real-time detection of surface defects in manufacturing processes of aluminum alloy strip using a lightweight network architecture. *J. Intell. Manuf.* **2022**. [CrossRef]
30. Wang, T.; Su, J.; Xu, C.; Zhang, Y. An Intelligent Method for Detecting Surface Defects in Aluminium Profiles Based on the Improved YOLOv5 Algorithm. *Electronics* **2022**, *11*, 2304. [CrossRef]
31. Yang, Y.; Sun, Q.; Zhang, D.; Shao, L.; Song, X.; Li, X. Improved Method Based on Faster R-CNN Network Optimization for Small Target Surface Defects Detection of Aluminum Profile. In Proceedings of the 2021 IEEE 15th International Conference on Electronic Measurement & Instruments (ICEMI), Nanjing, China, 29–31 October 2021; pp. 465–470.
32. Li, B.; Ren, F.; Ni, H.; Kang, X.; Lv, S.; Hao, Z. Classification Method of Surface Defects of Aluminum Profile Based on Transfer Learning. In Proceedings of the 2022 International Conference on Machine Learning and Intelligent Systems Engineering (MLISE), Guangzhou, China, 5–7 August 2022; pp. 1–5.
33. Wu, D.; Shen, X.; Chen, L. Detection of Defects on Aluminum Profile Surface Based on Improved YOLO. In Proceedings of the 2022 Prognostics and Health Management Conference (PHM-2022 London), London, UK, 27–29 May 2022; pp. 468–472.
34. Mushtaq, Z.; Su, S. Environmental sound classification using a regularized deep convolutional neural network with data augmentation. *Appl. Acoust.* **2020**, *167*, 107389–107401. [CrossRef]
35. Kisantal, M.; Wojna, Z.; Murawski, J.; Naruniec, J.; Cho, K. Augmentation for small object detection. *arXiv* **2019**, arXiv:1902.07296.
36. Lin, L. An effective denoising method for images contaminated with mixed noise based on adaptive median filtering and wavelet threshold denoising. *J. Inf. Process. Syst.* **2018**, *14*, 539–551.
37. Simonyan, K.K.C.; Vedaldi, A.; Zisserman, A. Return of the devil in the details: Delving deep into convolutional nets. *arXiv* **2014**, arXiv:1405.3531.
38. Hu, J.; Shen, L.; Sun, G. Squeeze-and-excitation networks. In Proceedings of the IEEE Conference on Computer Vision and Pattern Recognition, Salt Lake City, UT, USA, 18–23 June 2018.
39. Hou, Q.; Zhou, D.; Feng, J. Coordinate attention for efficient mobile network design. In Proceedings of the IEEE/CVF Conference on Computer Vision and Pattern Recognition, Nashville, TN, USA, 20–25 June 2021.
40. Woo, S.; Park, J.; Lee, J.Y.; Kweon, I.S. Cbam: Convolutional block attention module. In Proceedings of the European Conference on Computer Vision (ECCV), Munich, Germany, 8–14 September 2018.

Disclaimer/Publisher's Note: The statements, opinions and data contained in all publications are solely those of the individual author(s) and contributor(s) and not of MDPI and/or the editor(s). MDPI and/or the editor(s) disclaim responsibility for any injury to people or property resulting from any ideas, methods, instructions or products referred to in the content.

Article

Study on Friction Characteristics of AA7075 Aluminum Alloy under Pulse Current-Assisted Hot Stamping

Jiansheng Xia [1,2,*], Rongtao Liu [2], Jun Zhao [1], Yingping Guan [1] and Shasha Dou [2]

1. Key Laboratory of Advanced Forging & Stamping Technology and Science, Ministry of Education of China, Yanshan University, Qinhuangdao 066004, China; zhaojun@ysu.edu.cn (J.Z.)
2. Yancheng Institute of Technology, College of Mechanical Engineering, Yancheng 224051, China
* Correspondence: xiajs@ycit.edu.cn; Tel.: +86-15861988970

Abstract: Friction during contact between metals can be very complex in pulse current-assisted forming. Based on stamping process characteristics, a reciprocating friction tester was designed to study the friction characteristics between AA7075 aluminum alloy and P20 steel under different current densities. Origin software was used to process the experimental data, and a current friction coefficient model was established for the pulse current densities. The results show that the friction coefficient of the aluminum alloy sheet decreased with the increase in the pulse current density (2–10 A/mm^2). After that, the friction mechanism was determined by observing microscopic morphology and SEM: some oxide cracked on the friction surface when the current was large. Finally, finite element simulations with Abaqus software and a cylindrical case validated the constant and current friction coefficient models. The thickness distribution patterns of the fixed friction coefficient and the current coefficient model were compared with an actual cylindrical drawing part. The results indicate that the new current friction model had a better fit than the fixed one. The simulation results are consistent with the actual verification results. The maximum thinning was at the corner of the stamping die, which improved the simulation accuracy by 7.31%. This indicates the effectiveness of the pulse current friction model.

Keywords: AA7075 aluminum alloy; friction; pulse current; friction coefficient; numerical simulation

1. Introduction

With the development of the automotive industry, lightweight materials have gradually replaced traditional steel and have become one of the hot spots in the development direction of the automotive industry [1–3]. Among them, aluminum alloy has become the first choice to replace traditional steel because of its low density, high strength, and good processing formability. It has received attention for its use in lightweight automobiles [4]. Among the different types, 7075 aluminum alloy has the best intensity and is commonly used in aircraft manufacturing [5]. Although the traditional hot forming technology can avoid the problems of easy cracking, it has a small drawing limit ratio and there is considerable rebound of the aluminum alloy at room temperature [6]. Moreover, the heating time is long, and the heating efficiency is low, which reduces the quality of the sheet metal forming parts. It is therefore urgent to find a new forming technology to replace the traditional forming technology [7]. Electro-assisted forming (EAF) [8–10] improves metal forming performance by applying pulse current-assisted metal forming. Many studies have shown [11] that deformation resistance and spring back can be reduced under a pulse current, improving the metal-forming accuracy and quality. Hence, pulse current-assisted metal-forming technology has gradually become a research focus in recent years. Lv Z et al. [12] studied the electro-plastic effect of drawing high-strength steel; their results showed that introducing a pulse current can effectively improve the forming performance of high-strength steel and increase its deep drawing depth. The friction between

sheet metal and stamping mold impacts the forming quality and simulation accuracy in current-assisted hot stamping technology. Therefore, studying the friction characteristics under different currents is essential to better apply current-assisted forming technology to the sheet metal-forming process.

In recent years, some researchers have established friction models from a macroscopic perspective to study the influence of process parameters on sheet forming. Nie Xin [13] and Tan Guang et al. [14] of Hunan University measured DP480 high-strength steel considering the influence of temperature on the friction coefficient and established a variable friction model, which showed that it could better describe the actual stamping situation. Wang Peng et al. of the Hunan University of Technology [15] considered the influence of different interface loads on the friction coefficient between aluminum alloy sheets and mold steel under boundary lubrication conditions through the pin friction testing machine. They established a variable friction model under different loads and through experimental verification and finite element simulation. They found that the error of the variable friction model was small and verified the effectiveness of the friction mode. Subsequently, Dou S et al. [16] established a mixed friction model considering the influence of the sliding velocity and boundary load on the friction coefficient under boundary lubrication conditions, and combined experimental verification with finite element simulation to verify the effectiveness of the hybrid friction model. X J Li et al. [17] used the anti-problem optimization method to explore the influence of friction behavior on forming quality. They tested hot stamping on 7075 aluminum alloy sheets with different lubricants and analyzed the effects of the force–displacement curve, surface morphology, and thickness distribution. The results showed that compared to the experimental results, the determined friction coefficient could accurately predict the force–displacement curve and thickness distribution of the formed parts under different lubrication conditions. Liu Yong [18], Xu Yupeng [19], and Li Jiahao [20] of the Wuhan University of Science and Technology conducted a detailed study on the high-efficiency stamping process of high-strength aluminum alloy, mainly including the high-temperature rheological behavior of sheets and the high-temperature friction and lubrication behavior of aluminum alloy and mold steel, and applied it to the simulation of aluminum alloy sheet hot stamping. They analyzed the rupture mechanism of aluminum alloy hot stamping forming and provided a basis and reference for the actual stamping.

In addition, other researchers have begun to explain friction behavior from a microscopic perspective, and friction models have been established based on this. C. Wang et al. [21] established a micromechanical friction model considering the influence of the temperature, contact pressure, volumetric strain, and relative sliding velocity on friction during the forming process, and verified the model by comparing the actual contact area and experimental results. The results showed that the model could be used for formability analysis and the prediction of optimal stamping parameters, providing theoretical guidance for actual stamping. However, C. Wang only established a cold stamping process friction model based on temperature-dependent micromechanics. Jenny Venem [22] established a multi-scale friction model considering local contact pressure, temperature, and strain. They then applied it to hot stamping. The results showed that the model could predict the friction of the actual stamping process well. Deng Liang et al. [23] proposed a finite element model of the friction process at the microscopic scale based on the high-temperature one-way friction experimental process. They analyzed the actual contact conditions under the microscopic mechanism. The results showed that undulating the contact surface caused the friction factor calculated using the finite element model of the friction process to change within the range of the set friction factor. J. Han [24] considered the changes in the tangential stiffness and friction coefficient caused by the difference in the stress distribution and established a modified stick–slip friction model. The experimental verification and finite element simulation showed that the model could describe the friction behavior of the contact parameters at different stages in the stick–slip process. Its simulation results agreed with the experimental data, which showed significant improvement in the prediction accuracy of the mechanical system's performance. The authors of [25] investigated the friction and occlusion properties of 7075 aluminum alloy sheets at different temperatures and found that the formation of a compaction layer on the wear surface affected the friction behavior. They established a friction evolution model, which showed that at 25 °C and 150 °C, the dominant friction mechanism was furrow friction, while at temperatures above 300 °C, the dominant friction mechanism was sticking. Similarly, the authors of [26], in conjunction with the hot forming of 7075 aluminum alloys, established a friction mechanism at high temperatures, loads, and sliding speeds, which were used to explain the friction mechanism and showed that the wear rate gradually decreased as the load and sliding speed increased.

With the general study of variable friction models, some researchers have begun to study friction models related to current factors. ZH. C [27] used the pin–disc friction testing machine and Matlab software to establish a mixed friction model based on friction, sliding velocity, and average load. They verified the effectiveness of the friction model by comparing it with the predicted value of the friction model through experimental verification. Afterwards, the team led by ZH. C [28] conducted experiments to study the relationship between frictional force and multiple current factors, established a LuGre static model for frictional force on a bowstring, introduced dynamic parameters to develop a LuGre dynamic model, and identified the model parameters using the genetic algorithm combined with simulation. The results demonstrated the superiority of the dynamic model, which can provide a reference for predicting frictional force and studying frictional wear performance. On this basis, Ping Yu [29] of Liaoning Technical University considered the law of change of fluctuating contact force and current with friction, combined this with the Stribeck friction model, established a modified hybrid friction model of contact force current, and verified the effectiveness of the friction model through experimental measurement and model prediction. However, the application of the above models has mainly been in the field of high-speed bullet trains. Jx Bao et al. [30] established a multi-scale friction model considering the effect of the current density and size from a microscopic perspective through a current-assisted compression test. The results showed that the model could predict the friction coefficient well.

Many researchers have studied the friction characteristics in the hot forming process, but the friction characteristics under a pulse current are unclear, and there is relatively little research on related content. In this study, we used AA7075–T6 aluminum alloy as the research object. We studied the influence of a pulse current on the friction coefficient of materials under dry friction conditions and established a current friction model. Finally, through simulation and experimental verification, we verified the effectiveness of the new friction model by comparing it with the predicted values of the fixed friction coefficient.

2. Materials and Methods

2.1. Experimental Materials

The sheet used for this friction test was an AA7075–T6 aluminum alloy sheet with a thickness of 0.5 mm, manufactured by Alcoa and marketed by Suzhou Xiehe Metal Co., Ltd. (Suzhou, China). The chemical composition is shown in Table 1. First, the aluminum alloy sheet was cut into 900 mm × 30 mm × 0.5 mm using the wire-cutting technique, as shown in Figure 1a. In addition, this test used P20 die steel as the friction sub, whose chemical composition is shown in Table 2. The die steel was heat-treated to a hardness of 50 HRC, and the structure dimensions are shown in Figure 1b. The samples were then placed in an ethanol solution, which was used to remove the oil from the surface of the samples. Finally, the samples were ultrasonically cleaned for 15 min and then sealed for storage.

Table 1. Chemical composition of AA7075 aluminum alloy (mass fraction, %).

Si	Ti	Cu	Mn	Mg	Cr	Zn	Fe	Al
0.06	0.04	1.56	0.02	2.36	0.19	5.69	0.15	Bal.

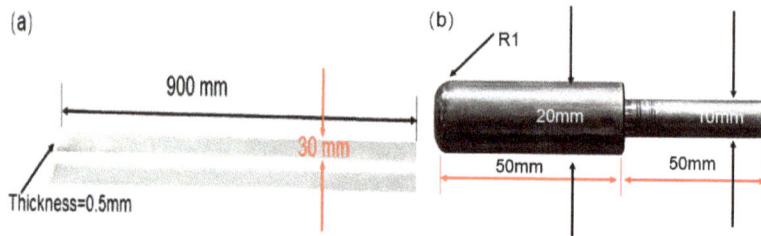

Figure 1. Structural dimensions of aluminum alloy strip and P20 steel pin. (**a**) 7075 Aluminum alloy strip. (**b**) P20 steel pin diagram.

Table 2. Chemical composition of P20 steel material (mass fraction, %).

C	Mn	Cr	Mo	S
0.38	1.3	1.85	0.40	0.008

The microscopic structure of the sample surface of the aluminum strip is shown in Figure 2a. This alloy predominantly comprises Al, Mg, and Zn, along with minor Fe and Si constituents. Due to the limited solubility of most alloying elements in Al, the microstructure of the alloy is characterized by a complex distribution of different particle phases over the α-Al solid solution matrix. The primary particle phases present in the 7075 alloys are η-$MgZn_2$, S-Al_2CuMg, T-$Al_2Zn_3Mg_2$, and T-$Al_2Zn_3Mg_2$. The non-equilibrium $MgZn_2$ phase is the primary strength-reinforcing phase. The complex phase particles are predominantly oriented along the tensile direction of the sample [31]. The EDS of this surface indicates that the main components of the original sample are Al, Mg, and Zn (Figure 2b).

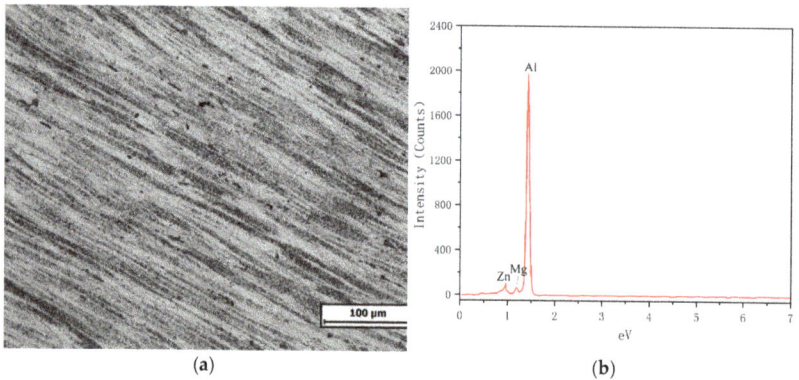

Figure 2. (**a**) A microscopic structure of the 7075 sample surface; (**b**) EDS surface analysis.

2.2. Test Principle

The pulse current friction tester, as shown in Figure 3a, mainly consisted of a friction test platform, control platform, pulse power loading platform, and data acquisition platform. The device can perform friction tests under the action of different currents, and its test schematic is shown in Figure 3b. In order to direct the current supplied by the external power supply to the friction test, the tester needs to be modified. The conductive clamps made by the group were installed on both sides of the aluminum alloy sheet, and the current was passed to the conductive clamps, P20 mold steel, and the sheet in turn to complete the closed circuit, as shown in Figure 3b.

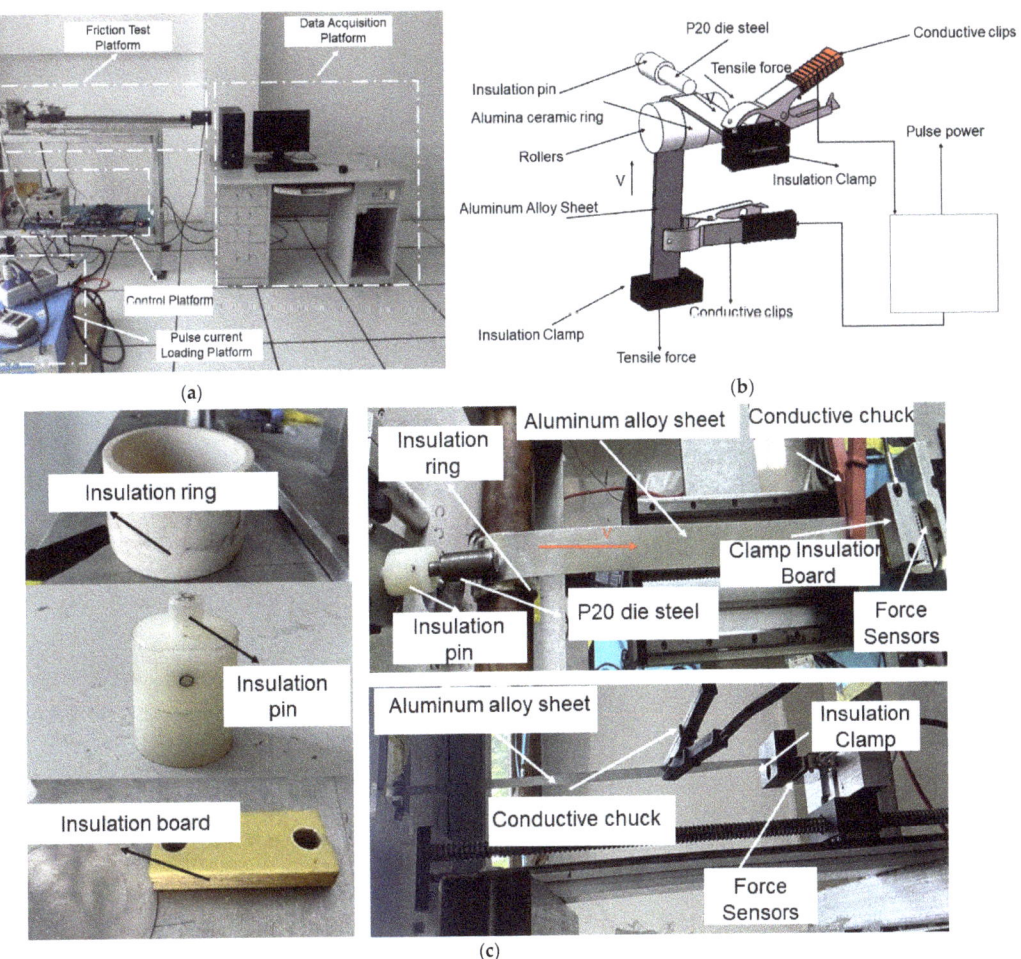

Figure 3. Testing machine. (**a**) Pulse current friction testing machine. (**b**) Schematic diagram of the friction testing machine. (**c**) Detailed parts of the friction test rig.

In addition, to ensure the safety of the test, an alumina ceramic insulation ring was installed on the drum, a nylon insulation pin was installed on the P20 mold steel, and an insulation spacer made of polyether ether ketone material was installed between the fixture and the plate material, thus insulating the conductive parts from the tester, as shown in Figure 3c.

The friction coefficient test platform included friction measurement components and vertical and horizontal actuators. The platform ensured insulation between the friction tester and the sheet material through the alumina ceramic ring, the epoxy resin sheet, and the nylon compression head. Using the fixtures to fix the two ends of the sheet, we connected to the two actuators and mounted the P20 die steel at the top surface of the sheet. The control platform comprised a stepping motor controller, driver, and motor power supply, which controlled the movement speed and direction of the two actuators and ensured the synchronous operation of vertical and horizontal directions. Two conductive clamps connected the pulse power platform to the sheet material. The pulse current was loaded through the conductive clamp, the roller, the material pressure head, and the sample material. It could quickly reach the temperature required for the friction test by adjusting the power supply parameters. The data acquisition system measured the force in the horizontal and vertical directions through the sensor installed in the two directions of the sheet and realized the automatic data acquisition with the help of LABVIEW programming. The formula of the friction coefficient was obtained using Coulomb's law of friction during data acquisition:

$$\mu = \frac{F}{2P} \qquad (1)$$

where, F is the difference between the horizontal and vertical sensor values, and P is the normal vertical load.

2.3. Test Arrangement and Test Procedure

The friction mechanism of aluminum alloys during pulsed current-assisted forming is complex, often accompanied by the coupled effects of electric and thermal fields and many influencing factors, such as the current density, temperature, normal load, sliding speed, etc. In this paper, we focus on the effect of a pulsed current on the friction characteristics of the contact interface between an aluminum alloy sheet and a P20 die steel. The specific test parameters are shown in Table 3.

Table 3. Specific test parameters for the friction test of 7075 aluminum alloys.

Load F_N (N)	Sliding Speed v (mm/s)	Current Density J (A/mm^2)	Stoke L (mm)	Lubricant
8	4	2, 4, 6, 8, 10	240	Dry friction

In this study, the steps in conducting the pulsed current friction test were as follows. First, the position of the plating fixture was adjusted so that the aluminum alloy sheet could be fixed to the test machine. After fixing the plates, conductive grips were installed at each end of the machine to ensure that the pulse power supply was energized. The plates were then preheated for 10 min using the pulsed power supply. After the preheating was complete, the current parameters were adjusted, and the plates were continuously energized. The temperature of the plates was monitored in real time using an infrared thermographer. Finally, the remaining test parameters were set using the computer program, and the friction test began. After the friction test reached the predetermined effective stroke, the friction test was completed. All friction tests were repeated three times to ensure the accuracy of the data.

2.4. Material Characterization Methods

The frictional wear mechanism of the 7075 aluminum alloy was analyzed by post-testing the specimens under the action of a pulsed current. Firstly, we utilized the VK-X100 laser scanning microscope to analyze its three-dimensional morphology after conducting the friction test on the aluminum alloy sheet. Secondly, the wear surfaces of the plates were then analyzed using the JXA-840A scanning electron microscope (SEM) and the energy dispersive spectrum (EDS), and then the chemical composition was analyzed. Finally, the shape of the aluminum alloy ports perpendicular to the sliding direction of the aluminum alloy was also analyzed using SEM.

3. Results

3.1. Friction Coefficient at Different Current Densities

The aluminum alloy sheet produced a Joule heating effect when the pulse current was loaded, and the temperature of the sheet metal increased, which is significant in explaining the friction characteristics of metal sheet forming. The Optris infrared thermometer was used to measure the temperature in real time and the curve of the sheet metal temperature with the time under different current densities was obtained using Origin software, as shown in Figure 4. As seen in the Figure, the temperature curves under different current densities had a standard feature. The heating rate of the aluminum alloy sheet metal was fast until 20 s and tended to be stable with the continuous increase in time. The main reason is that the Joule thermal temperature under the pulse current increased with the influence of the air convection and thermal radiation coefficient on the sheet at the beginning of the test [32–34].

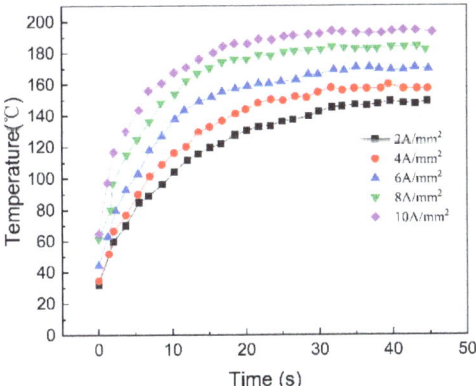

Figure 4. Curve of sheet metal temperature with time under different current densities.

The current intensity values of the frictional tests were 30 A, 60 A, 90 A, 120 A, and 150 A, and the corresponding current densities were 2 A/mm^2, 4 A/mm^2, 6 A/mm^2, 8 A/mm^2, and 10 A/mm^2, respectively. Under dry friction conditions, the friction coefficient curved with time at different current densities with a load of 8N and a speed of 4 mm/s (Figure 5a). In the graph, the coefficients of friction at different current densities have common characteristics: the coefficients of friction increased rapidly at first, then gradually decreased, and then finally plateaued. Moreover, the friction coefficient gradually decreased with the increase in the current density.

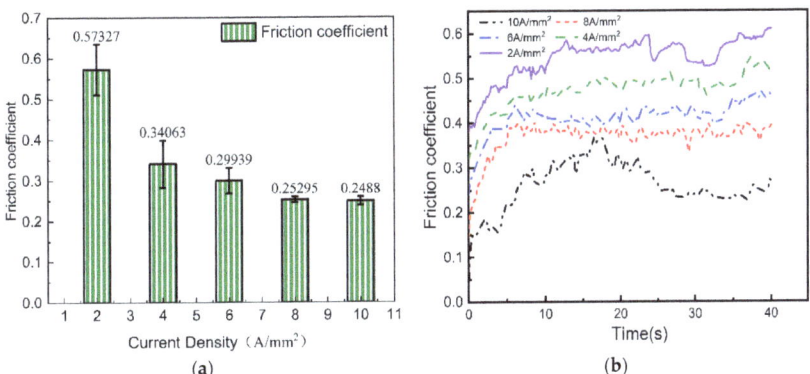

Figure 5. (a) Friction coefficient curves with time at different current densities. (b) Average coefficient of friction in the stabilization phase.

The effects of the current on the coefficients of friction are mainly manifested in the following aspects:

(1) Current conduction in the material is limited to the conductive filaments. Due to the current's contraction effect, it accumulates at the conductive filament, resulting in a rapid increase in temperature, which causes softening of the microprotrusions. This, in turn, reduces the biting depth and shear degree between the two contact surfaces, leading to less damage and significantly reduced roughness components at the contact interface.

(2) The passage of current through the subsurface of the plate has a specific softening effect, which reduces plastic deformation. This softening effect is believed to enhance the migration of subsurface electrons, resulting in a decrease in the friction coefficient.

(3) The increase in temperature caused by the current flow can lead to the rapid formation of an oxide film on the surface of the friction pair, which can act as a lubricant. As the current increases, the oxidation process accelerates, resulting in the formation of more lubricating oxide films on the contact surface of the aluminum alloy, which provides protection against wear. As a result, the friction coefficient tends to decrease with the increasing current, as the oxide film formed by oxidation serves as an effective lubricant [35,36].

The average coefficients of friction with different current densities are shown in Figure 5b. The friction coefficients vary with different current densities as follows: $(\mu_{2}=0.573) > (\mu_{4}=0.341) > (\mu_{6}=0.299) > (\mu_{8}=0.253) > (\mu_{10}=0.249)$. The maximum friction coefficient was 0.573 when the current density was 2 A/mm^2, and the minimum friction coefficient occurred at 10 A/mm^2. Although the tests were carried out in the dry friction state, the average friction coefficients were lower than about 0.5, which was close to the friction coefficient in the boundary lubrication state. This phenomenon is normal, and the authors of [30] mention that proper current density can increase lubrication and thus reduce mechanical wear.

3.2. The Effect of Current Density on Surface Friction Mechanism

After the friction test, a VK-X100 laser microscope was used to observe the surface morphology of the AA7075 aluminum alloy at different currents, as shown in Figure 6. When the current density was 2 A/mm^2, few scratches and peeling occurred on the aluminum alloy sheet surface (Figure 6a). A possible reason is the instantaneous heating effect of the pulse current, softening the metal. Under the dual impacts of Joule heat and frictional heat, the viscosity of the material increased, resulting in a higher friction coefficient value, and the primary friction mechanisms were viscous wear and furrow wear. When the current density increased to 4 A/mm^2, the peeling traces on the sheet material disappeared, but it had more dents and deep scratches (Figure 6b). A possible reason is that the current density increased the continuous action of Joule heat, the metal surface oxide film reduced the friction coefficient, and the friction mechanism was furrow wear. When the current increased to 6 A/mm^2, the number of dents decreased gradually, and the width became shallow (Figure 6c). A possible reason is the shrinkage effect of the current: the temperature of the pulse current at the conductive spot increased, resulting in the softening of the micro-convex body, thus reducing the friction coefficient, and the primary friction mechanism was furrow wear. As shown in Figure 6d, when the current reached 8 A/mm^2, there were still multiple dents on the board, but the width and depth of the marks were reduced. The surface was relatively smooth, and the critical friction mechanism was furrow wear. When the current continued to rise to 10 A/mm^2, there were only a few dents on the board, the friction marks became shallow, and the friction mechanism was furrow wear, as shown in Figure 6e.

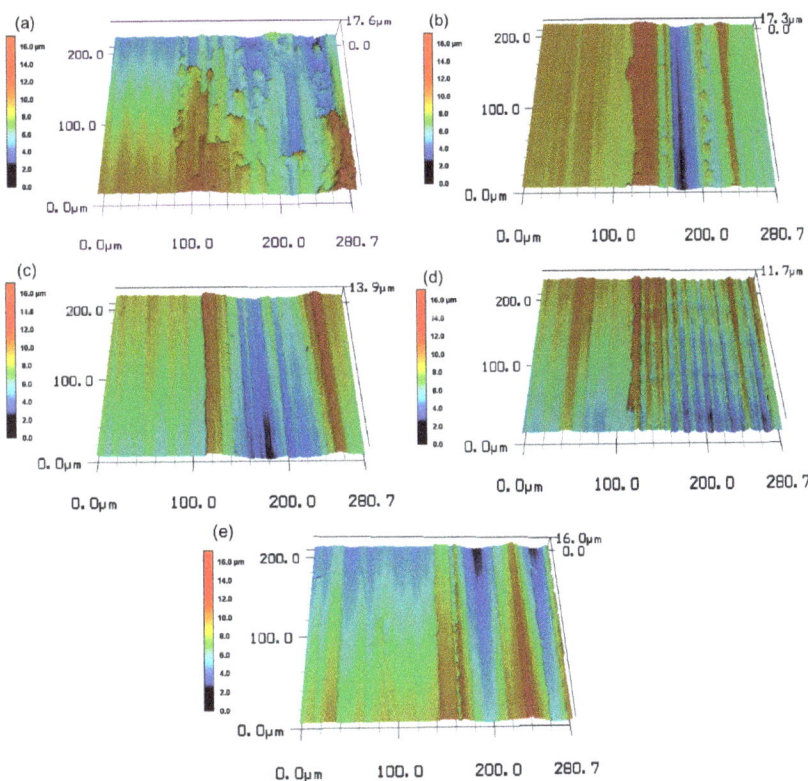

Figure 6. Surface morphology of 7075 aluminum alloy sheet at different current densities. (**a**) 2 A·mm^{-2}; (**b**) 4 A·mm^{-2}; (**c**) 6 A·mm^{-2}; (**d**) 8 A·mm^{-2}; (**e**) 10 A·mm^{-2}.

To further investigate the friction and wear mechanism under a pulsed current, the microscopic morphology of the surface of the aluminum alloy sheet was observed via a scanning electron microscopy (SEM) with a pulsed current density of 6 A·mm^{-2} as an example, and the results are shown in Figure 7. As shown in Figure 7a, wear marks parallel to the sliding direction were observed on the friction specimen, while the wear surface was relatively smooth with only shallow grooves present. In addition, some oxidation cracks were observed in the morphology beneath the wear surface, as shown in Figure 7b. The chemical composition of the wear surface of the sheet was further analyzed using EDS, and the results in Figure 8 show that it was mainly Al and O, with the mass fraction of O reaching 37.3% (i.e., 80% of the overall composition of the aluminum oxide). The generation of a continuous oxide film on the friction surface effectively avoided direct metal-to-metal contact between the friction pairs and reduced the occurrence of adhesive wear. This is mainly because when the pulsed current was passed into the metal during the process, the current easily produced an oxide layer on the aluminum alloy surface, and the current improved the oxidation properties of the sliding surface through surface polarity. The Joule heating effect of the pulsed current on the metal and the heat from the friction process also made the aluminum alloy surface more susceptible to oxidation, while the friction coefficient gradually decreased as a result.

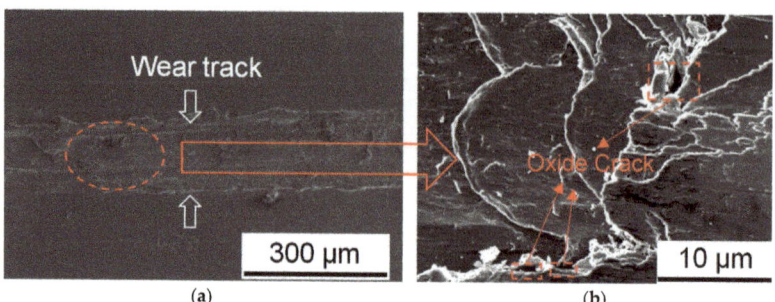

Figure 7. SEM diagram of the frictional profile of an aluminum alloy sheet at a current density of 6 A·mm^{-2} at different magnifications; (**a**) Surface profile perpendicular to the wear track (**b**) Oxide cracks under surface morphology.

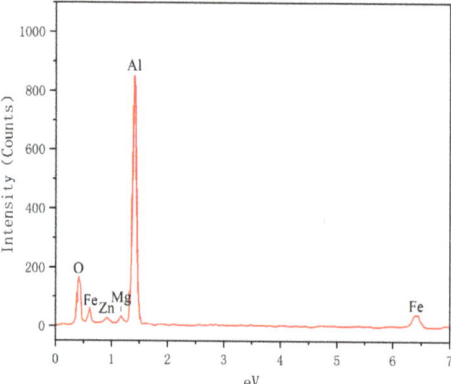

Figure 8. EDS results of the surface of a sheet at a current density of 6 A/mm^2.

To investigate the frictional wear mechanism under pulsed current, the cross-sectional specimens of the aluminum alloy were observed using scanning electron microscopy (SEM) after the friction test, as depicted in Figure 9. A high-temperature softening effect occurs in aluminum alloy under the influence of the current, as seen from the microscopic state of the surface morphology of the cross-section after friction. The white layer that appears in the middle of the cross-sectional after friction, as shown in Figure 9a, is due to the frictional heat generated during the friction process and the Joule heat after the introduction of the current. This results in adhesive deformation on the upper surface of the aluminum alloy, leading to an increase in the frictional area and surface roughness of the sheet. The morphology of the intermediate white layer was further observed in Figure 9b, which revealed the presence of a mechanically mixed layer (MML) formed by the combination of the oxide generated during the wear process and the substrate under the action of the pressure of the two contact surfaces. The composition of the MML layer was analyzed using EDS, and the results in Table 4 showed that at the wear of 6 A·mm^{-2}, the O content was 18.1%, the Al content was 67.8%, and the Fe content was 10.9%. The presence of Fe indicated the occurrence of material transfer on the wear surface of the plate, where the mold material was attached to the surface of the aluminum alloy sheet [37].

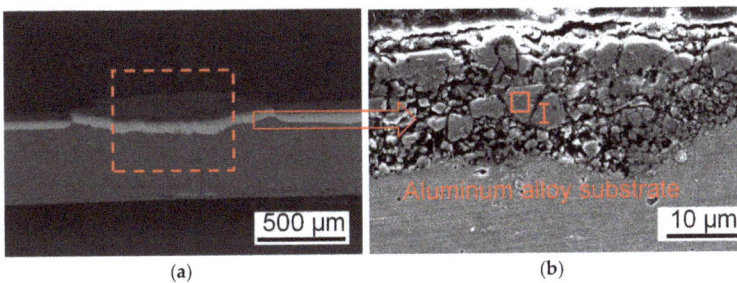

Figure 9. SEM surface morphology of frictional wear cross-sections of aluminum alloys at different magnifications at a current density of 6 A/mm²; (**a**) Section shape of the aluminum alloy sheet after wear; (**b**) Mechanically mixed layer underwear morphology.

Table 4. EDS results of worn cross-section of the 7075 alloy at 6 A/mm² (wt.%).

O	Al	Fe
18.1%	67.8%	10.9%

3.3. Establishment of Friction Coefficient Model Based on the Current Density

As can be seen in Figure 5b, when the current density was less than 6 A·mm⁻², the average friction coefficient decreased with the increase in the current density. However, the trend of the friction coefficient decreasing with the increasing current density became stable when the current density was more than 6 A·mm⁻². One reason is that before the current density increased to 6 A·mm⁻², the Joule heat and frictional heat generated by instantaneous heating during the pulse current reduced the friction coefficient. The other reason is the low sliding speed: the micro-convex body between the surface of the plate and the surface of the P20 steel had enough time to produce plastic deformation, which increased the contact area and ultimately led to an increase in the friction coefficient of the plate. So, the friction coefficient decreased with the increasing current density and stabilized when the current density was above 6 A·mm⁻². According to the above analysis and the law of curve change, the pattern of the evolution of the friction coefficient with the current density conforms to an inverse function. Therefore, the new friction coefficient expression is as follows:

$$\mu = \frac{a}{J+b} + c \tag{2}$$

where, μ is the coefficient of friction; J is the current density; and a, b, and c are constants. Then, the friction data were imported into the Origin software to fit the inverse function curve shown in Figure 10. The error between the fitted curve and the experimental data was small, and the fitting degree was 0.993, so the function accurately reflects the current friction coefficient with the current density. Using the Origin software fit, $a = 0.50$, $b = -0.68$, and $c = 0.19$ were obtained. Thus, the new friction model expression is as follows:

$$\mu = \frac{0.50}{J - 0.68} + 0.19 \tag{3}$$

Five groups of different current densities (1 A/mm², 3 A/mm², 5 A/mm², 7 A/mm², and 9 A/mm²) were selected to verify the correctness of the new current friction model. Five sets of actual experimental measurements under different current densities and the predicted values of the new friction model are shown in Table 5. As can be seen, the total errors were less than 9%, the lowest error was 3.89%, and the maximum error was 8.13%. This indicates that the fitted current friction model can better reflect the law of the friction coefficient with different current densities in the metal forming, verifying the effectiveness of the current friction model.

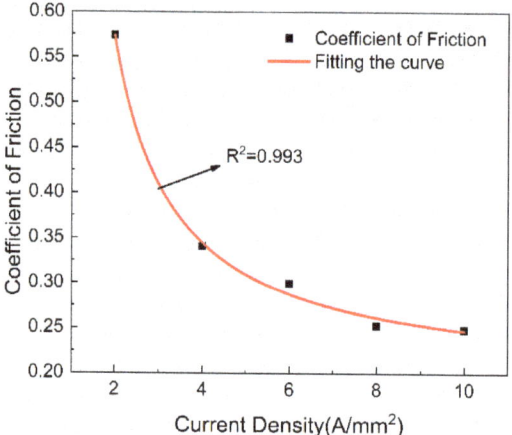

Figure 10. Friction fitting curves at different current densities.

Table 5. Friction coefficient measurement values and friction model prediction values.

Current Density (A/mm^2)	1	3	5	7	9
Friction coefficient measurement values	0.878	0.412	0.315	0.319	0.277
Current friction model prediction values	0.840	0.439	0.339	0.293	0.266
Error (%)	4.25	6.57	7.39	8.13	3.89

4. Simulation and Experimental Validation

To improve the accuracy of the software simulation and verify the accuracy of the new friction model at different current densities, we adopted the sequential coupling model for simulation, used the same AA7075 aluminum alloy as the research object for the thermoelectric coupling simulation, and then imported the simulation temperature field results into the thermal stamping as a predefined field. Then, we intuitively and effectively analyzed the temperature, strain, and stress fields under the different process parameters to verify the effectiveness of the new current friction model.

4.1. Finite Element Analysis of Thermoelectric Coupling

In the thermoelectric coupling simulation, the copper electrode and the AA7075 aluminum alloy sheet models improved the computational efficiency, as shown in Figure 10. The aluminum alloy sheet had a diameter of 165 mm and a thickness of 0.5 mm; to facilitate current loading, the ends on both sides had a length of 60 mm and a width of 30 mm. The size of the grid mesh was 1 mm, the unit size was 4 mm, and the grid number was 29,696. The size of the copper electrode was 120 mm × 120 mm × 60 mm, the unit size was 2.5 mm, and the grid number was 4396. Due to the thermoelectric coupling analysis involving the current field, temperature field, potential field, etc., the mesh type of the model was set as an 8-node linear hexahedral thermoelectric coupling element (DC3D8E).

In thermoelectric coupling, the pulse current will bring the Joule heating effect. Hence, copper electrodes and aluminum alloys use temperature-related parameters, citing the simulation-related data of [38], as shown in Tables 6 and 7.

Table 6. Thermal performance parameters of the copper electrode at different temperatures.

Temperature (K)	Density (kg/m³)	Specific Heat Capacity (J/kg·K)	Thermal Conductivity (W/m·K)	Electroconductibility (Ω⁻¹/mm⁻¹)
298	8930	0.385	400	59,170
373	8890	0.397	395	
473	8850	0.408	388	34,130
573	8800	0.419	382	
673	8740	0.427	376	24,814
713	8690	0.434	370	

Table 7. Thermal performance parameters of aluminum alloys at different temperatures data from [38].

Temperature (K)	Density (kg/m³)	Specific Heat Capacity (J/kg·K)	Thermal Conductivity (W/m·K)	Thermal Expansion (μm/m·k)	Electroconductibility (Ω⁻¹/mm⁻¹)
298		0.85	121.1	21.6	
373		0.90	129.4	23.4	
473	2810	0.95	138.6	23.6	19,417
573		0.97	146.6	24.3	
673		1.00	154.1	25.2	
713		1.08	160.5	25.6	

In the thermoelectric coupling simulation, the heat transfer coefficient between the sheet and the surrounding environment is cited in the literature [39], as shown in Table 8. In the temperature boundary, the entire model was set to a room temperature of 298.15 K through a predefined field. The current was loaded on the sheet metal through the node set of the copper electrode. A zero-potential boundary was set on the other side of the sheet. The contact between the aluminum alloy metal sheet and the copper electrode was set to normal hard contact and a penalty function: when the distance between the aluminum alloy sheet and the copper electrode is less than 0.1 mm, the contact thermal conductivity is 10 mW/m²·K.

Table 8. Conduction heat transfer coefficient of aluminum alloy surface and air.

Temperature (K)	373	473	573	673	773
Convection heat exchange coefficient (W/m²·K)	17	18	19	22	23

4.2. Finite Element Analysis of Thermal Stamping Forming

SolidWorks software was used to model the punch, die, and blank holder, and then the data were imported into ABAQUS after assembly, as shown in Figure 11a. The essential dimensions of the mold are shown in Table 9. The meshing type was the thermoelectric structure coupling unit (Q3D8R), the unit size was 2 mm, and the total number of grids was 11,132. The size of the sheet material was consistent with the thermoelectric coupling simulation, and the grid type was the thermoelectric structure coupling unit (Q3D8R). The analysis step units used thermoelectric structure coupled units, and the meshing assembly model is shown in Figure 12b.

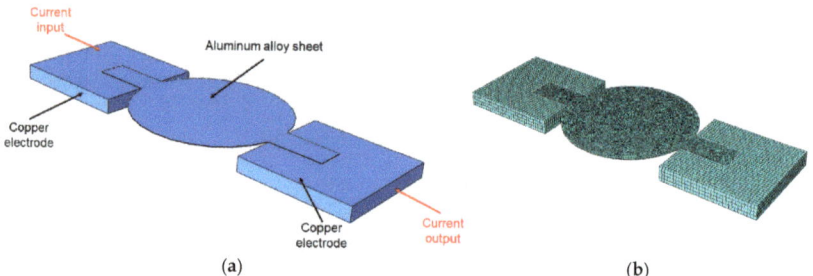

Figure 11. Thermoelectric-coupled finite-element simulation model. (a) Thermoelectric coupling finite element simulation model. (b) Hot stamping finite element simulation model.

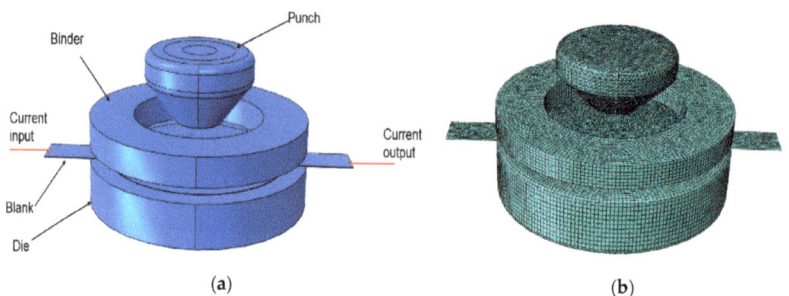

Figure 12. Finite element meshing diagram. (a) Thermoelectric coupling finite element meshing. (b) Current-assisted hot stamping finite element meshing.

Table 9. Specific parameters of stamping die (unit length: mm).

Punch Profile Radius (mm)	Bottom Round Angle Radius (mm)	Die Profile Radius (mm)	Pressure Rim Diameter (mm)	Die Profile Radius (mm)	Die Profile Radius (mm)
20	4	60	120	106	5

The mold material was P20 stainless steel, with a density of 7.81 g/cm³, Young's modulus of 205 GPa, and Poisson's ratio of 0.275, and its thermal performance parameters are shown in Table 10. The sheet metal material was also an AA7075 aluminum alloy, and its material parameters were consistent with the parameters in the thermoelectric coupling simulation. The stress–strain curve data are cited [39]. In the temperature boundary, the temperature field results of the thermoelectric coupling simulation were imported as the initial conditions into the predefined fields of the thermoelectric-structure coupling. The current was loaded on the sheet metal through the node. The current direction is shown in Figure 11a. Zero-potential loading occurred on the other side of the sheet. The die was set as a fixed constraint, the vertical displacement of the punch was set to 35 mm, and the pressing force was set to 2.5 kN. When the distance between the sheet and the mold was less than or equal to 0.1 mm, the contact thermal conductivity was 10 W/m²·K. When the space was more than 0.1 mm, the value was 0.

Table 10. Thermal properties of P20 steel.

Density (g/cm³)	Yield Stress (MPa)	Heat Conductivity (W/m·K)	Thermal Expansion (10^{-6} °C)	Specific Heat Capacity (J/kg·K)	Electroconductivity (Ω^{-1}/mm^{-1})
7.81	836	31.5	12.8	460	1370

4.3. Results Analysis and Example Verification

The sheet temperature field distribution at a current density of 10 A·mm^{-2} for 100 s is shown in Figure 13. In Figure 13, the highest temperature is about 450 K, and the lowest temperature in the middle of the sheet is 319 K. The high temperature is in the middle area, in which the distribution is relatively uniform, the temperature at both ends is low, and the change is not constant. The process considers the convection heat transfer coefficient of aluminum alloy sheet and air, the temperature rise of the joule thermal effect, and the contact between copper electrode and the left and right ends, which is why the temperature of aluminum alloy material was relatively high near the middle. The temperature change of the contact electrodes was small because the resistivity of the copper electrode was less than that of the aluminum alloy sheet.

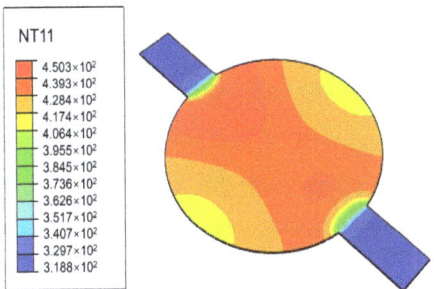

Figure 13. Temperature field distribution of sheet material after thermoelectric coupling simulation.

The electro-assisted thermal stamping simulation was carried out in two groups: one used a fixed friction value, $\mu = 0.1$, and the other used the new current friction model. The new friction model was written in the Fortran language with a subroutine of "fri-coef" to submit jobs to the Job module. Figure 14a,b show the equivalent plastic strain after thermal stamping simulation under the different friction models. As can be seen in the graphs, the equivalent peak plastic strain at the fixed friction values was greater than the current friction coefficient model, and the maximum equivalent plastic strain was concentrated at the rounded corners of the punch. Because the sheet metal was subjected to the stretching effect at the bottom of the punch fillet during the deep drawing process, the sheet metal had high fluidity.

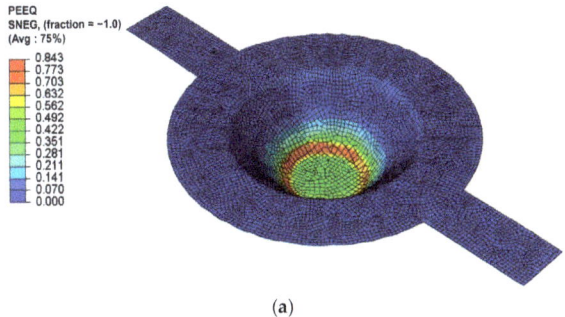

(a)

Figure 14. Cont.

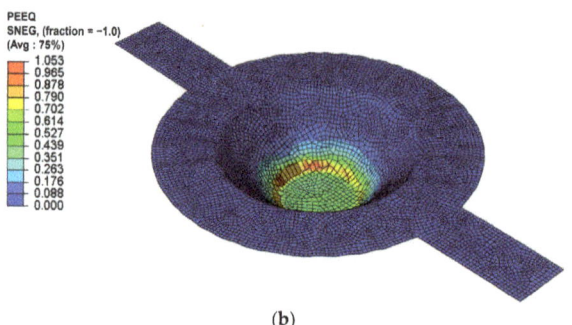

(b)

Figure 14. The equivalent plastic strain after thermal stamping simulation under different friction models. (**a**) The fixed friction value of $\mu = 0.1$. (**b**) The new current friction model.

The sheet material was a cylindrical part with a radius of 165 mm and a thickness of 0.5 mm. The stamping speed was 20 mm/s, and the pressure edge force was consistent with the simulation, set to 2.5 kN. Figure 15a shows the actual electrical-assistance stamping platform, Figure 15b shows the stamped part, and Figure 15c shows the locations of the measurement points. The 3D model extracted the node coordinates through the path to observe the thickness rule, using the shell unit to study the thickness distribution and thinning rate. Origin software obtained the thickness distribution patterns (Figure 15d). Compared to the traditional fixed friction coefficient, the new current friction model was closer to the thickness rule of the actual stamping parts. The maximum thinning rate of the fixed friction coefficient was 12.5%, the current friction coefficient model was 15.3%, and the actual value was 14.3%. Compared to the actual value, the simulation error of the software decreased from 12.59% to 6.99%, and the simulation accuracy improved by 7.31%. The maximum thinning was also concentrated at the circular corner of the punch, consistent with the simulation results of the maximum equivalent plastic strain and within the allowable range of thinning. From the thickness distribution curves of the part, the current friction model was closer to the actual value. Therefore, the current friction model reflects the friction characteristics of sheet forming and provides a practical reference for current-assisted stamping simulation technology.

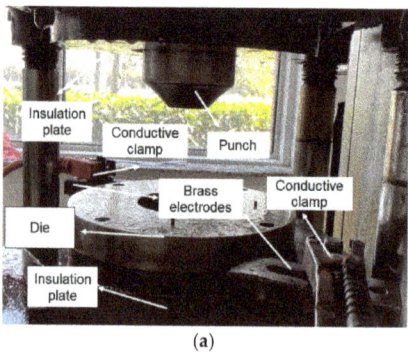

(a)

(b)

Figure 15. *Cont.*

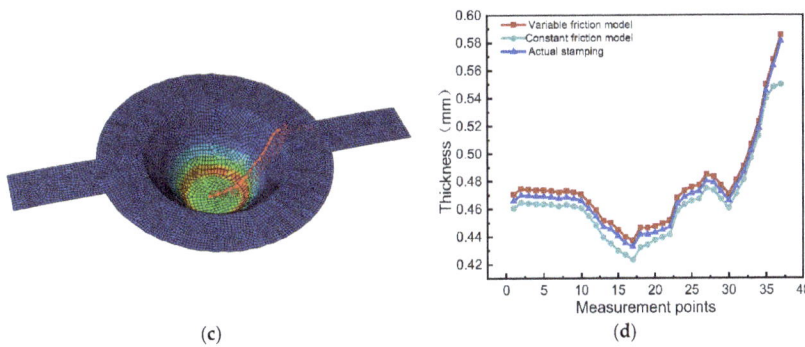

Figure 15. (**a**) Current-assisted stamping platform. (**b**) Actual stamped part. (**c**) Thickness measurement point. (**d**) The thickness distribution of different friction models.

5. Conclusions

(1) With a speed of 4 mm/s and under the conditions of dry friction, the friction coefficient between the AA7075 aluminum alloy sheet and the P20 die steel gradually decreased with the increase in the current density, and the decreasing trend gradually slowed down. According to the surface morphology, there were many scratches and a small amount of peeling on the AA7075–T6 aluminum alloy surface, and the primary friction mechanism was furrow wear and viscous wear.

(2) In pulse current-assisted stamping forming, tears occurred on the friction surface, and oxides appeared when the current increased, mainly caused by the electric heating effect.

(3) When the current density was 2–10 $A \cdot mm^{-2}$, based on the variable friction model established at the current density, the fitting degree was good, which can accurately describe the friction behavior of the AA7075 aluminum alloy and the P20 die steel.

(4) In pulse current-assisted stamping forming, the maximum thinning of the aluminum alloy plates occurred at the rounded corners of the convex shape, and the maximum thinning rate of the variable friction coefficient was greater than that of the constant friction coefficient. The maximum equivalent plastic strain was also concentrated at the corner of the punch, corresponding to the thinning rate result. The thickness distribution of the variable friction coefficient model was more consistent with the actual stamping parts, which verifies the effectiveness of the current friction model.

Author Contributions: Conceptualization, J.X. and J.Z.; Methodology, J.X. and Y.G.; Software, S.D. and R.L.; Validation: J.X., J.Z. and Y.G.; Data collection, R.L. and J.X.; Data analysis, R.L. and S.D.; Writing—original draft preparation, R.L. and J.X.; Writing—review and editing, J.X. and J.Z.; Supervision, J.Z. and Y.G. All authors have read and agreed to the published version of the manuscript.

Funding: This research was funded by China's National Natural Science Foundation [51505408].

Institutional Review Board Statement: Not applicable.

Informed Consent Statement: Not applicable.

Data Availability Statement: Not applicable.

Conflicts of Interest: The authors declare no conflict of interest.

References

1. Li, G.; Liu, X. Overview of the research status of automotive lightweight technology. *Mater. Sci. Technol.* **2020**, *28*, 47–61. [CrossRef]
2. Wang, S.; Wang, D.; Huang, Y.; Meng, Q.; Zhang, L.; Dong, X. Research status of aluminum alloy material application in the field of automotive lightweight. *Alum. Process.* **2022**, 3–6.
3. Ren, L.; Xu, Y. Application and development of metal Materials in automobile Lightweight. *Times Automob.* **2021**, *352*, 38–39.

4. Liu, Y.; Geng, H.; Zhu, B.; Wang, Y.; Zhang, Y. Research progress of high efficiency hot stamping process of high strength aluminum alloy. *Forg. Technol.* **2020**, *45*, 1–12. [CrossRef]
5. Guo, Y.; Xie, Y.; Wang, D.; Zhao, J.; Du, L. Constitutive Model and Process Analysis of 2124 Aluminum Alloy Hot Forming. *Forg. Technol.* **2022**, *47*, 213–219. [CrossRef]
6. Xia, J.; Zhao, J.; Dou, S. Friction Characteristics Analysis of Symmetric Aluminum Alloy Parts in Warm Forming Process. *Symmetry* **2022**, *14*, 166. [CrossRef]
7. Nie, D.; Lu, Z.; Zhang, K. Hot V-bending behavior of pre-deformed pure titanium sheet assisted by electrical heating. *Int. J. Adv. Manuf. Technol.* **2018**, *94*, 163–174. [CrossRef]
8. Dong, H.-R.; Li, X.-Q.; Li, Y.; Wang, Y.-H.; Wang, H.-B.; Peng, X.-Y.; Li, D.-S. A review of electrically assisted heat treatment and forming of aluminum alloy sheet. *Int. J. Adv. Manuf. Technol.* **2022**, *120*, 7079–7099. [CrossRef]
9. Tiwari, J.; Balaji, V.; Krishnaswamy, H.; Amirthalingam, M. Dislocation density based modelling of electrically assisted deformation process by finite element approach. *Int. J. Mech. Sci.* **2022**, *227*, 107433. [CrossRef]
10. Xiao, A.; Huang, C.; Yan, Z.; Cui, X.; Wang, S. Improved forming capability of 7075 aluminum alloy using electrically assisted electromagnetic forming. *Mater. Charact.* **2022**, *183*, 111615. [CrossRef]
11. Lv, Z.; Zhou, Y.; Zhan, L.; Zang, Z.; Zhou, B.; Qin, S. Electrically assisted deep drawing on high-strength steel sheet. *Int. J. Adv. Manuf. Technol.* **2021**, *112*, 763–773. [CrossRef]
12. Nie, X.; Xiao, B.; Shen, D.; Guo, W. Thermodynamically coupled stamping studies considering the deformation heat and frictional thermal effects. *China Mech. Eng.* **2020**, *31*, 2005–2015. [CrossRef]
13. Tan, G. *Study on Cold Stamping Forming of High Strength Steel Based on Heat-Force Coupling and Variable Friction Coefficient*; Hunan University: Changsha, China, 2017.
14. Wang, P. *Experimental Study and Finite Element Analysis of 5052 Aluminum Alloy Sheet Stamping Forming Friction Characteristics*; Hunan University of Technology: Zhuzhou, China, 2018.
15. Dou, S.; Xia, J. Analysis of Sheet Metal Forming (Stamping Process): A Study of the Variable Friction Coefficient on 5052 Aluminum Alloy. *Metals* **2019**, *9*, 853. [CrossRef]
16. Li, X.; Yan, X.; Zhang, Z.; Ren, M.; Jia, H. Determination of Hot Stamping Friction Coefficient of 7075 Aluminum. *Metals* **2021**, *11*, 1111. [CrossRef]
17. Liu, Y. *Study on High Efficiency Hot Stamping Process and High Temperature Rheological and Friction Behavior of High Strength Aluminum Alloy Sheet*; Huazhong University of Science and Technology: Wuhan, China, 2018.
18. Xu, Y. *Study on Hot Stamping Forming Friction and Lubrication Behavior of 7075 Aluminum Alloy Sheet Material*; Huazhong University of Science and Technology: Wuhan, China, 2019.
19. Li, J. *Study on the Adhesive Friction and Wear Behavior of 7075 Aluminum Alloy during Hot Stamping*; Huazhong University of Science and Technology: Wuhan, China, 2021.
20. Wang, C.; Hazrati, J.; De Rooij, M.B.; Veldhuis, M.; Aha, B.; Georgiou, E.; Drees, D.; Van den Boogaard, A.H. Temperature dependent micromechanics-based friction model for cold stamping processes. *J. Phys. Conf. Ser.* **2018**, *1063*, 012136. [CrossRef]
21. Venema, J.; Hazrati, J.; Atzema, E.; Matthews, D.; van den BOOGAARD, T. Multi-scale friction model for hot sheet metal forming. *Friction* **2022**, *10*, 316–334. [CrossRef]
22. Deng, L. Finite element simulation of p-friction process of plate and die at mesoscopic scale. *J. Plast. Eng.* **2022**, *29*, 196–201. [CrossRef]
23. Han, J.; Ding, J.; Wu, H.; Yan, S. Mechanism analysis and improved model for stick-slip friction behavior considering stress distribution variation of interface. *Chin. Phys. B* **2022**, *31*, 034601. [CrossRef]
24. Lu, J.; Song, Y.; Hua, L.; Zhou, P.; Xie, G. Effect of temperature on friction and galling behavior of 7075 aluminum alloy sheet based on ball-on-plate sliding test. *Tribol. Int.* **2019**, *140*, 105872. [CrossRef]
25. Lu, J.; Song, Y.; Zhou, P.; Lin, J.; Dean, T.A.; Liu, Y. Process parameters effect on high-temperature friction and galling characteristics of AA7075 sheets. *Mater. Manuf. Process.* **2021**, *36*, 967–978. [CrossRef]
26. Chen, Z.; Sun, G.; Shi, G.; Li, C. Study on characterization and model of friction of sliding electrical contact of pantograph-catenary system. In Proceedings of the IECON 2017—43rd Annual Conference of the IEEE Industrial Electronics Society, Beijing, China, 29 October–1 November 2017; IEEE: Piscataway, NJ, USA; pp. 2312–2317. [CrossRef]
27. Chen, C.H.; Jia, L.; Shi, G.; Hui, L.; Tang, J. A dynamic friction model for bow network carrying flow under fluctuating load. *J. Liaoning Univ. Eng. Technol. (Nat. Sci. Ed.)* **2021**, *40*, 48–55. [CrossRef]
28. Ping, Y. *Study on Current-Carrying Friction Characteristics and Friction Force Modeling in Bow Net under Fluctuating Contact Force*; Liaoning Engineering and Technical University: Fuxin, China, 2019.
29. Bao, J.; Bai, J.; Lv, S.; Shan, D.; Guo, B.; Xu, J. Interactive effects of specimen size and current density on tribological behavior of electrically-assisted micro-forming in TC4 titanium alloy. *Tribol. Int.* **2020**, *151*, 106457. [CrossRef]
30. Ge, L.; Wang, S.; Yang, Z. Wear behavior of 7075 aluminum alloy and its mechanism. *Spec. Cast. Non-Ferr. Alloy.* **2011**, *31*, 178–182+205. [CrossRef]
31. Guo, F.; Chen, M.; Chen, Z.; Shi, G.; Hui, L. Study on friction characteristics and modeling of sliding electric contact. *J. Electrotech. Technol.* **2018**, *33*, 2982–2990. [CrossRef]
32. Li, X.; Xu, Z.; Guo, P.; Peng, L.; Lai, X. Electroplasticity mechanism study based on dislocation behavior of Al6061 in tensile process. *J. Alloy Compd.* **2022**, *910*, 164890. [CrossRef]

33. Liu, Y.; Wan, M.; Meng, B. Multiscale modeling of coupling mechanisms in electrically assisted deformation of ultrathin sheets: An example on a nickel-based superalloy. *Int. J. Mach. Tools Manuf.* **2021**, *162*, 103689. [CrossRef]
34. Ugurchiev, U.K.; Novikova, N.N. Features of the Analytical Method to Determine the Temperature during Electroplastic Rolling. *J. Mach. Manuf. Reliab.* **2023**, *51* (Suppl. S1), S79–S83. [CrossRef]
35. Chen, Z.H.; Dang, W.; Shi, G.; Wang, X.L.; Liu, F.S. Modelling of friction in sliding electrical contacts under fluctuating loads. *J. Electr. Eng. Technol.* **2019**, *34*, 5126–5134. [CrossRef]
36. Chen, Z.H.; Tang, J.; Shi, G.; Hui, L.C.; Jia, L.M. Analysis and modeling of frictional vibration of strong current sliding electrical contacts in bow networks. *J. Electr. Eng. Technol.* **2020**, *35*, 3869–3877. [CrossRef]
37. Brown, L.; Xu, D.; Ravi-Chandar, K.; Satapathy, S. Coefficient of Friction Measurement in the Presence of High Current Density. *IEEE Trans. Magn.* **2007**, *43*, 334–337. [CrossRef]
38. Lu, J. *7075 Basic Research of Key Technology of Hot Stamping of High Strength Aluminum Alloy*; Wuhan University of Technology: Wuhan, China, 2021.
39. Gu, R.; Liu, Q.; Chen, S.; Wang, W.; Wei, X. Study on High-Temperature Mechanical Properties and Forming Limit Diagram of 7075 Aluminum Alloy Sheet in Hot Stamping. *J. Mater. Eng. Perform.* **2019**, *28*, 7259–7272. [CrossRef]

Disclaimer/Publisher's Note: The statements, opinions and data contained in all publications are solely those of the individual author(s) and contributor(s) and not of MDPI and/or the editor(s). MDPI and/or the editor(s) disclaim responsibility for any injury to people or property resulting from any ideas, methods, instructions or products referred to in the content.

Article

Modified Voce-Type Constitutive Model on Solid Solution State 7050 Aluminum Alloy during Warm Compression Process

Haihao Teng [1,2], Yufeng Xia [1,2,*], Chenghai Pan [1,2] and Yan Li [1,2]

1. School of Material Science and Engineering, Chongqing University, Chongqing 400044, China; liyan11@cqu.edu.cn (Y.L.)
2. Chongqing Key Laboratory of Advanced Mold Intelligent Manufacturing, College of Materials Science and Engineering, Chongqing University, Chongqing 400044, China
* Correspondence: yufengxia@cqu.edu.cn

Abstract: The 7050 alloy is a kind of Al-Zn-Mg-Cu alloy that is widely used for aircraft structures. Although the deformation behavior of the solid solution state 7050 aluminum alloy is critical for engineering and manufacturing design, it has received little attention. In this study, the room and warm compression behavior of the solid solution-state 7050 alloy was researched, and a modified model with variable parameters was built for the flow stress and load prediction. The isothermal compression tests of the solid solution-state 7050 alloy were performed under the conditions of a deformation temperature of 333–523 K, a strain rate of 10^{-3}–10^{-1} s^{-1}, and a total reduction of 50%. The strain-stress curves at different temperatures were corrected by considering interface friction. The flow stress model of aluminum was established using the modified Voce model. For evaluating the modified Voce model's prediction accuracy, the flow stresses calculated by the model were compared with the experimental values. Consequently, for assessing its prediction abilities in finite element applications, the whole compression process was simulated in the finite element analysis platform. The results sufficiently illustrated that the modified Voce-type model can precisely predict the complex flow behaviors during warm compression. This study will guide the prediction of the warm compression load and the optimization of the heat treatment process of the alloy.

Keywords: solid solution state; 7050 aluminum alloy; warm compression; constitutive model

Citation: Teng, H.; Xia, Y.; Pan, C.; Li, Y. Modified Voce-Type Constitutive Model on Solid Solution State 7050 Aluminum Alloy during Warm Compression Process. *Metals* **2023**, *13*, 989. https://doi.org/10.3390/met13050989

Academic Editor: Elisabetta Gariboldi

Received: 22 March 2023
Revised: 1 May 2023
Accepted: 9 May 2023
Published: 19 May 2023

Copyright: © 2023 by the authors. Licensee MDPI, Basel, Switzerland. This article is an open access article distributed under the terms and conditions of the Creative Commons Attribution (CC BY) license (https://creativecommons.org/licenses/by/4.0/).

1. Introduction

The high-strength, age-hardened 7050 alloy is a kind of Al-Zn-Mg-Cu alloy that is widely used for aircraft structures. After the forging of the 7050 alloy, it will be heated to a solid solution temperature and kept for some time [1]. Then the alloy will be quenched and finally artificially aged for improved performance [2,3]. The treatment will induce large residual stresses on the structures, resulting in severe distortion and even failure in the subsequent machining stage [4,5].

In previous studies, the warm compression between solid solution and aging treatment has been proven to have a good residual stress reduction effect on large aluminum alloy forgings [6,7]. The load of the warm compression process is large due to the high strength of the solid solution-state 7050 alloy [8]. To predict the load precisely, the deformation behavior and applicative constitutive model of solid solution-state 7050 aluminum at medium temperature are essential, but they are little studied. Some constitutive models, such as Swift, Ludwik, and Voce, have been used to predict flow stresses and loads in cold or warm forming designs [9–11]. At low temperatures, there is also softening due to mechanical work converting. However, the indicated constitutive models ignore softening phenomena in the deformation and are insufficient to predict flow stresses and loads during warm compression for the alloy.

Therefore, the goal of this research is to build a model of 7050 aluminum alloy in its solid solution state for flow stress and load prediction during warm compression. Room temperature and isothermal warm compress tests from 333 to 523 K were performed. The flow properties of the aluminum in its solid solution state were studied. By considering the friction compensation, the modified Voce-type model with parameters variable by the effects of strain, temperature, and strain rate was established for flow stress prediction. For evaluating the modified model's prediction accuracy, the flow stresses calculated by the model were compared with the experimental values. The model's prediction abilities in finite element applications have also been assessed.

2. Materials and Methods

The 7050 aluminum alloy for the research was from Deyang Wanhang Die Forging Co., Ltd. (Deyang, China). The chemical compositions of the aluminum alloy are listed in Table 1.

Table 1. Chemical composition of 7050 alloy (wt, %).

Cr	Cu	Fe	Mg	Mn	Si	Ti	Zn	Al
0.04	2.0	0.16	1.90	0.10	0.126	0.06	5.7	Bal

The solid solution treatment and compressing processes are shown in Figure 1. The wrought 7050 alloy was heated and solution treated at 750 K for 4 h with 333 K water quenching. The quenched alloys were prepared into some cylinders of $\Phi 8 \times 12$ mm for subsequent compressive tests. The warm upsetting tests were conducted on a Gleeble−3180 unit at different temperatures of 333, 423, 473, and 523 K for strain rates of 10^{-3}, 10^{-2}, and 10^{-1} s^{-1}. The total reduction is 50%, with stress-strain data recorded automatically. Before upsetting, tantalum with graphite lubricant was applied to the surfaces of both specimens and dies to minimize the friction effect. The specimens were heated at 10 K/s to deformation temperatures and then held for some time to eliminate the temperature gradient and microstructure inhomogeneity. The room-temperature compression tests of forged and solid solution states were carried out on a WDW-100 universal tester (Beijing Sinofound Co., Ltd, Beijing, China). The strain-stress curves are recorded by the tester automatically.

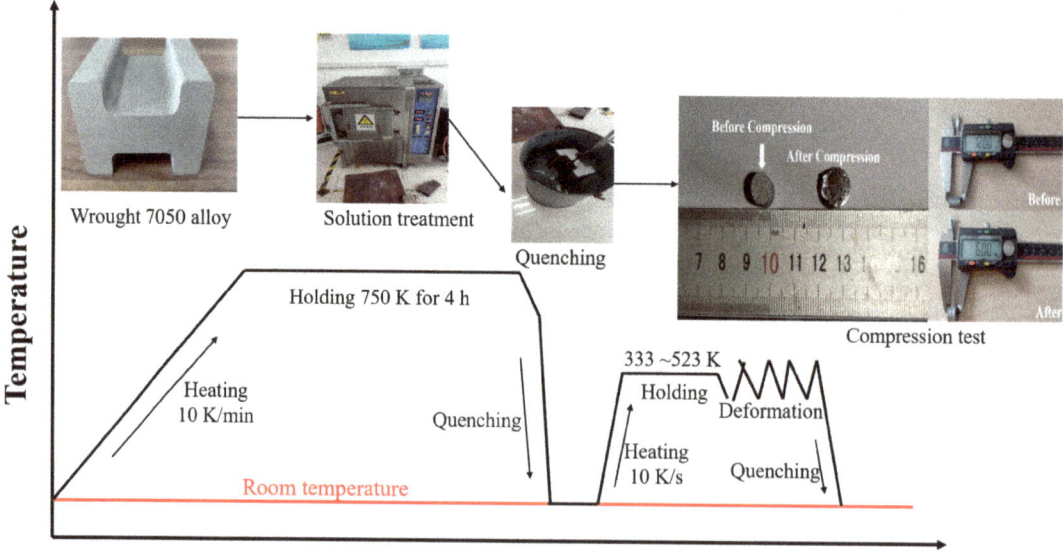

Figure 1. Schematic diagram of the experimental process for the 7050 alloy.

The samples before and after heat treatment were also wire-electrode cut and metallographically observed for comparison. The samples were ground and polished to eliminate any trace of cutting. Then they were etched with Graff-Sargent solutions. Optical micrographs of the samples in different states were characterized by an OLYMPUS GX-41 type microscope(Olympus Corporation, Tokyo, Japan). The microstructure was further examined using a TESCAN VEGA3 LMH scanning electron microscope (TESCAN CHINA Ltd., Shanghai, China).

3. Results and Discussion
3.1. Flow Behavior

The stress-strain curves of the 7050 alloy at room temperature in three states (forged state, solid solution state, and annealed state after forging) were obtained and depicted in Figure 2. From the comparison, it can be concluded that the solid solution-stated alloy has greater strength, plasticity, and fracture toughness. Firstly, as shown in Figure 3a–c, the grains in the wrought state were fibrous. After quenching, the grain is equiaxial (shown in Figure 3d–f), indicating that the treatment enhances recrystallization grain generation [12–14]. The volume fraction of recrystallized grains of 7050 alloy before and after solid solution treatment is measured by the software Image-Pro-Plus 6.0 (Media Cybernetics Inc., Rockville, MD, USA). The values are 18.9% and 52.6%, respectively. The volume fraction of recrystallized grains of 7050 alloy measured in other experiments is shown in Figure 4 [15–17]. The treatment can enhance the recrystallization of the 7050 aluminum. The recrystallization reduces intragranular dislocation density and intragranular-grain boundary strength differences, which will decrease work-hardening and increase plasticity. Secondly, the supersaturated solid solution generated in the treatment will finally transfer to the strengthening phase (the white region of the SEM image in Figure 5b), which will pin dislocation and improve the alloy strength [18]. Thirdly, the coarse second-phase particles of the alloy (the black part in Figure 5a) lessened during the treatment [19]. The coarse particles (above 2 nm) of 7050 alloy decreased by about 70% after solid solution treatment [20]. The refined particles hinder void generation and crack propagation to enhance the fracture toughness of the 7050 alloy.

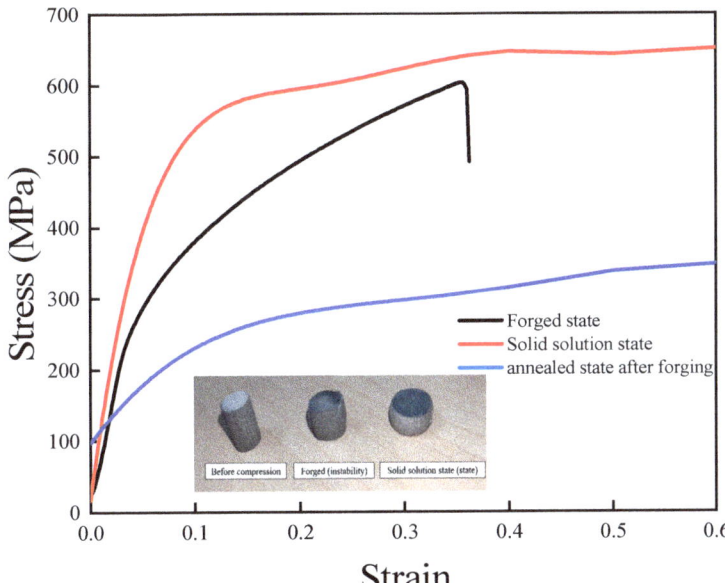

Figure 2. Stress-strain curve for room temperature compression of 7050 alloys in different states.

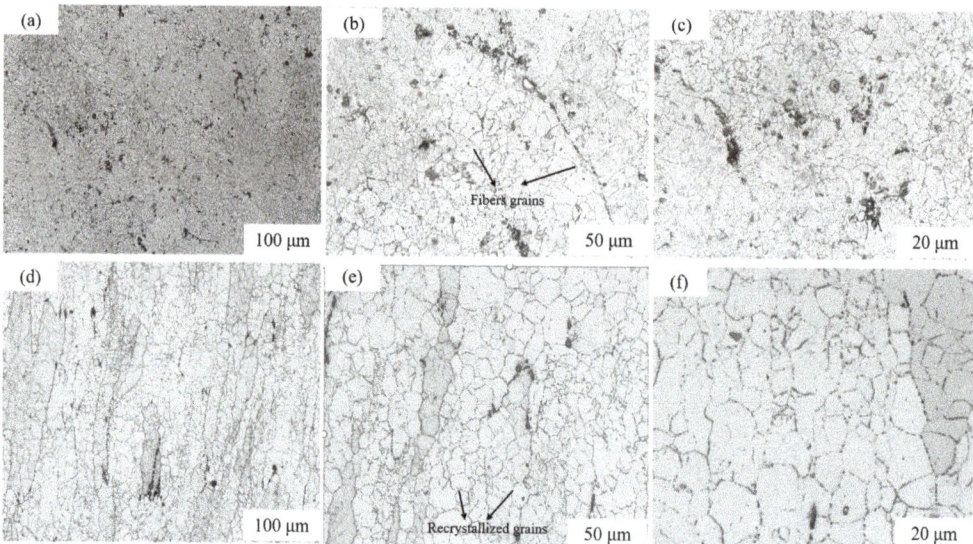

Figure 3. Optical micrographs of 7050-aluminum alloy before and after solid solution treatment (**a–c**) Before solid solution treatment; (**d–f**) After solid solution treatment.

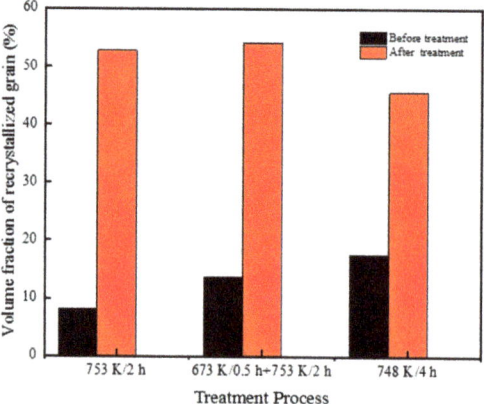

Figure 4. The volume fraction of recrystallized grains before and after solid solution treatment [15–17].

The true stress-strain curves of aluminum in the solid solution state at elevated temperatures from 333 K to 523 K at different strain rates are shown in Figure 6a–c. All curves could be divided into three distinct stages. In the first stage, the flow stress increases linearly with strain, and only elastic deformation predominates. In the second stage, plastic deformation happens and flow stress slowly increases. In the third stage, the flow stress will approach saturation and weave. Furthermore, it is seen that flow stress is negatively correlated with deformation temperature. The microstructure evolution (precipitation or recrystallization), deform-mechanism conversion (shear mechanism to Orowan mechanism), and reduction of dislocation movement resistance occurring at warmer temperatures can all alter the flow characteristics and decrease flow stress [21–23]. At lower temperatures, only the strain-hardening tendency exhibits itself in the curve. When the deformation temperature is approximately 423 K, the work hardening occurs first, and then the 7050 alloy begins to soften at a higher plastic strain. The tendency for flow softening becomes more obvious at warmer deformation temperatures. Moreover, the flow stress of solid solution-state 7050 alloys has a positive correlation with strain rate. At high strain rates, dislocation density and dislocation slip velocity increase, resulting in a magnification of dislocation interaction and deformation resistance [24,25]. Additionally, at the same compression temperature, the amplification of flow stress with increasing strain rate is roughly the same. It is due to the balance of work hardening and deformation thermal softening at medium temperatures [26]. The strain rate effect of solid solution 7050 aluminum alloy is not obvious during the medium-temperature compression process, and the stress-strain curve is not sensitive to the strain rate [27].

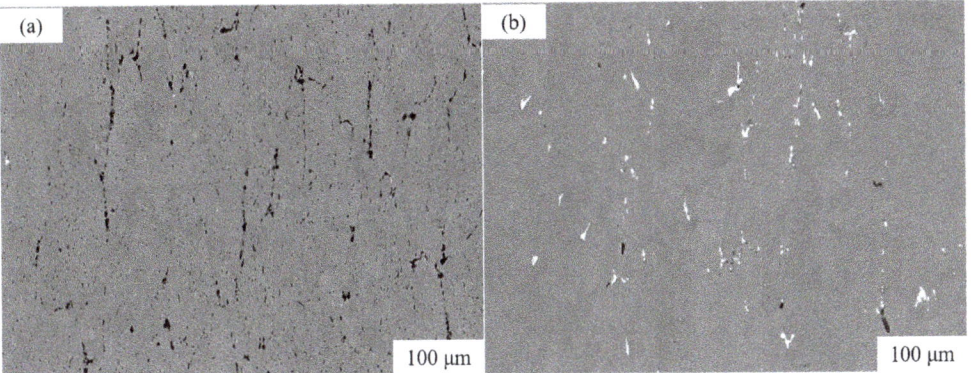

Figure 5. SEM images of the 7050 alloy (**a**) without heat treatment; (**b**) after quenching.

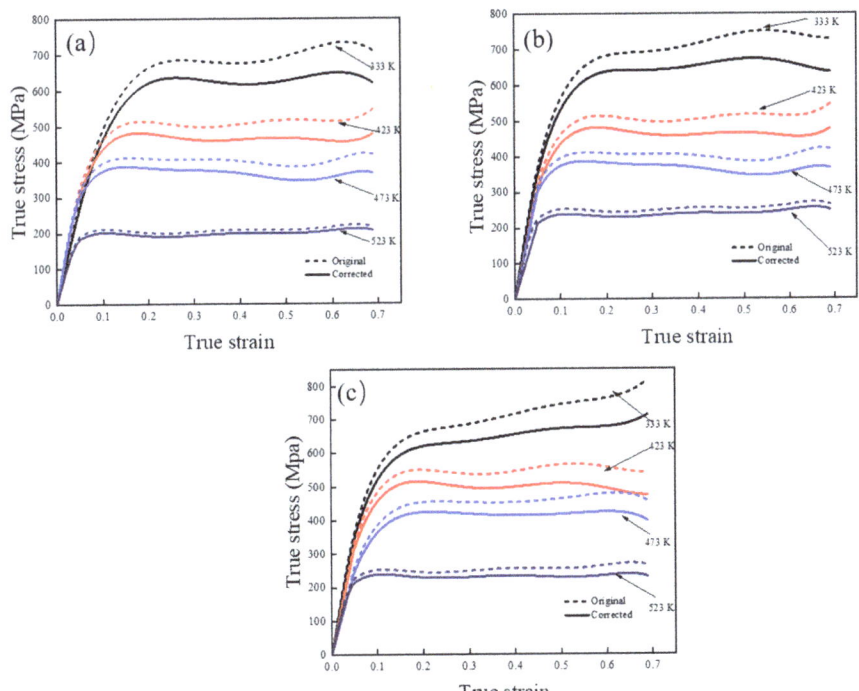

Figure 6. The flow stress of solid solution-state 7050 alloy in different conditions: (**a**) 0.001 s^{-1}; (**b**) 0.01 s^{-1} (**c**) 0.1 s^{-1}.

In compression tests, the friction will change the stress state, leading to heterogeneous deformation and non-negligible errors in the obtained flow stress, although the necessary lubricant was applied [28]. Friction correction is necessary for the accurate calculation of flow stress. The correction method is expressed in Equation (1) [29,30]:

$$\sigma = \frac{\overline{\sigma}}{1 + \left(\frac{2}{3\sqrt{3}}\right) m \left(\frac{R_0}{h_0}\right) \exp\left(\frac{3\varepsilon}{2}\right)} \quad (1)$$

where σ is the corrected stress, $\overline{\sigma}$ is the measured stress without correction, and ε is the strain, respectively. For this specimen, R_0 and h_0 respectively, represent the initial radius and height of the compressed sample (unit: mm). The R_0-value is 4 and the h_0-value is 12 for this compression. Additionally, m is the friction factor for warm upsetting, and the value is 0.2. Compared with the flow stress with and without correction, the measured values are significantly higher than the actual ones. The deviation increases with strain due to the incremental contact area. Also, the work-hardening effects are reduced after correction.

3.2. Construction and Comparison of Constitutive Models

The flow stress in the room and at medium temperature can be analyzed using Hollomon [31], Swift [32], Ludwigson [33], and Voce hardening models [34]. All constitutive models are shown in Equations (2)–(4) below:

$$\text{Hollomon } \sigma = k_H \varepsilon^{n_H} \quad (2)$$

$$\text{Swift } \sigma = k_S (\varepsilon 0 + \varepsilon)^{n_{Swift}} \quad (3)$$

$$\text{Ludwigson } \sigma = k_1 \varepsilon^n + e^{k_2} e^{n_2 \varepsilon} \quad (4)$$

$$\text{Voce } \sigma = \sigma_0 \left(1 - A e^{-k\varepsilon}\right) - k\varepsilon \quad (5)$$

where σ is the true stress in the compression, ε is the true strain, n is the strain hardening index, σ_0, ε_0, k and A are material constants.

Besides, some classical viscoplastic models, such as the Arrhenius-type model [35–37] and the Johnson–Cook model [38], were applied for comparison. The two models are shown in the equation as follows:

$$\text{Arrhenius-type model } \dot{\varepsilon} = A[\sinh(\alpha\sigma)]^n \exp\left(-\frac{Q}{RT}\right) \quad (6)$$

where α, and n are material constants. Q is the activation energy of plastic deformation (J/mol); R is the gas constant; and T is the temperature.

$$\text{Johnson–Cook model } \sigma = (A + B\sigma^n)\left(1 + C \ln \frac{\dot{\varepsilon}}{\dot{\varepsilon}_{ref}}\right)\left(1 - T_H^D\right) \quad (7)$$

where A, B, and C are material constants. $\dot{\varepsilon}_{ref}$ is the reference strain rate; T_H is the relative temperature.

In the above equations, all the parameters are obtained by the reference or fitted by the Levenberg-Marquardt approach. The predictions of the flow stress at 423 K and 0.1 s^{-1} by the four models are shown and compared in Figure 7. The fitting degrees of the four models are different, and the order of priority is as follows: Voce > Swift > Ludwigson > Hollmon > Johnson-Cook > Arrhenius. The Arrhenius-type model uses a hyperbolic sine function to predict the flow stress of the material, and it cannot accurately predict the hardening behavior of the alloy at medium temperature. The unsaturated models (Johnson-Cook, Swift, Ludwigson, and Hollmon) utilize a certain exponent to predict the flow stress with strain hardening. The flow stress calculated by the models was greatly less than the measurement before the peak point. Due to the identical hardening exponent in the models, flow stress in the middle and late stages of compression will be exaggerated. [39]. However, for the saturation models (Voce model), it can better fit the different hardening rates at different states with variable exponents, and it can also better predict the saturated flow stress at stable states. For solid solution-state 7050 alloy during warm compression, the flow stress has a saturated value after the hardening stage, and the voice-type model is best for prediction.

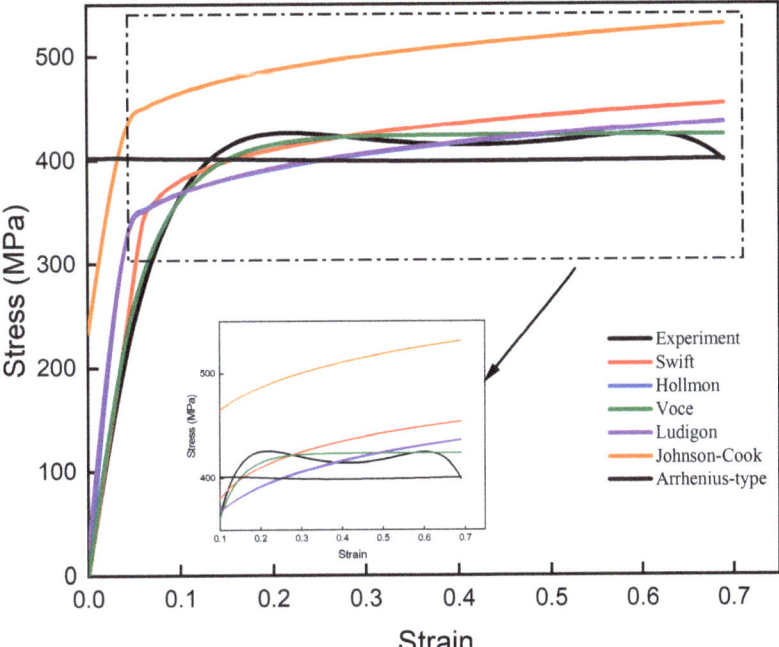

Figure 7. Fitting results of the flow stress at 423 K in 0.01 s^{-1}.

3.2.1. Voce-Type with Softening Coefficient

The original Voce model ignores the softening in high strain and the effects of temperature and strain rate on flow stress. To improve accuracy, a Voce-type with a hardening and softening coefficient is employed to describe the deformation behaviors with saturated flow stress at room and medium temperatures as follows [40,41]:

$$\begin{cases} \sigma = k\dot{\varepsilon}^{m_0+m_1T} \\ k = (1 - k_{Sof})k_{Har} \end{cases} \quad (8)$$

$$\begin{cases} k_{Har} = K(\varepsilon + \varepsilon_0)^n \exp(\frac{\beta}{T}) \\ K_{Sof} = 1 - \exp(-(r_0 + r_1 T)\varepsilon) \end{cases} \quad (9)$$

where σ is the flow stress (unit: MPa), T is the absolute temperature (unit: K), k_{Har} is the strain strengthening index, and K_{Sof} stands for the softening index. Besides, β, m_0, m_1, r_0, and r_1 are dimensionless material constants.

Both sides of Equations (8) and (9) are taken as the natural logarithm and changed to Equation (10);

$$\ln \sigma = \ln k + n\ln(\varepsilon + \varepsilon_0) + \frac{\beta}{T} - (r_0 + r_1)\varepsilon + (m_0 + m_1T)\ln \dot{\varepsilon} \quad (10)$$

All the material constants in the Voce model can be determined from the corrected strain-stress curves. At a given temperature and strain, hardening and softening coefficients are fixed. Therefore, Equation (8) can be simplified as follows:

$$\ln \sigma = (m_0 + m_1 T) \ln \dot{\varepsilon} + k_2 \quad (11)$$

Therefore, $m_0 + m_1 T$ is the slope of $\ln\sigma$-$\ln\dot{\varepsilon}$ plot shown in Figure 8a. Then, the slopes in different conditions are shown in Figure 8b. The m_0-value can be determined from the slope of plots, and the m_1-value is obtained from the plots' intercept in Figure 8b.

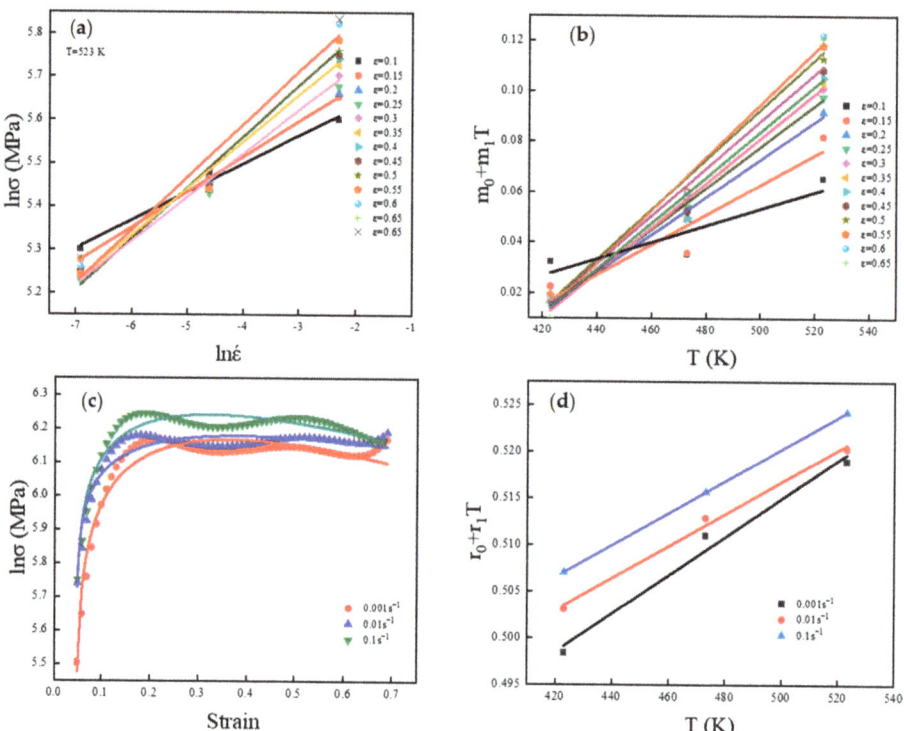

Figure 8. Parameters fitting under different conditions: (a) $\ln\sigma$-$\ln\dot{\varepsilon}$ (b) $m_0+m_1 T$-T (c) $\ln\sigma$-strain (d) $r_0 + r_1 T$-T.

At a given temperature, $r_0+r_1 T$ is a constant and is set to k_3. $\ln k + n \ln (\varepsilon + \varepsilon_0) + (m_0 + m_1 T) \ln \dot{\varepsilon}$ is also a constant at a certain temperature and the strain rate and should be set to K_4. Equation (11) is changed to Equation (12);

$$\ln \sigma = n \ln(\varepsilon + \varepsilon_0) - k_3 \varepsilon + k_4 \quad (12)$$

Figure 8c shows the relation between $\ln\sigma$ and ε, and Equation (12) is employed to characterize the relationship. The fitting result reveals the value of the material constants in different conditions. Besides, k_3-T plots are shown in Figure 8d. The slopes and intercepts of the plots were calculated to determine the r_0-value and r_1-value in the model. Similarly, the constants K and β can be obtained from the k_4-T plots. All the material constants in the voce-type model are obtained, and the constitutive model is shown as follows:

$$\sigma = 533.0765\left((\varepsilon - 0.02593)^{0.2842}\right) \times \mathrm{EXP}\left(\frac{142.13587}{T}\right) \times \mathrm{EXP}(-(0.43411 + 0.000162213T)\varepsilon) \; \dot{\varepsilon}^{0.00683 + 0.0000183178T} \quad (13)$$

3.2.2. Modified Voce-Type Model with Variable Parameters

The material parameters in the conventional voce-type model are invariable. However, they are influenced by temperature and other conditions in the warm deformation due to microstructure evolution, which decreases simulation accuracy. The modification to the constitutive model is that the wave of material parameters can be corrected. The modified constitutive model with variable material parameters includes two parts: the Voce constant prediction and parameter wave correction. The prediction of the Voce constant (F_{Voce}) is based on the traditional model in Equations (8) and (9). Then, a corrected function K_c is set as a compensation for the parameter wave in warm deformation. Therefore, Equation (8) can be rewritten as follows:

$$\begin{cases} \sigma = K_c\, F_{Voce} \\ F_{Voce} = (\varepsilon + \varepsilon_0)^n \exp(\frac{\beta}{T}) \exp(-(r_0 + r_1 T)\varepsilon)\, \dot{\varepsilon}^{(m_0 + m_1 T)} \\ K_c = k_T K_\varepsilon K_{\dot{\varepsilon}} \end{cases} \quad (14)$$

where σ is the flow stress (unit: MPa), T is the absolute temperature, and m_0, m_1, r_0, r_1, ε_0 are dimensionless material constants. k_T, K_ε and $K_{\dot{\varepsilon}}$ are the compensation coefficients that can be used to describe the effects of temperature, strain, and strain rate on flow stress respectively.

In the isothermal compression process, the deformation heat raises the alloy's temperature and affects the deformation behavior [42,43]. Therefore, the coupling effect of strain and deformation temperature on material parameter evolution must be considered. Meanwhile, the flow stress of 7050 alloy at warm temperatures is not sensitive to the strain rate due to the balance between work hardening and deformation thermal softening. The interaction between strain rate and deformation temperature and material parameter evolution could be neglected, and the compensation coefficients K_c can be described as follows:

$$K_c = k_{T,\varepsilon} K_{\dot{\varepsilon}} \quad (15)$$

where $k_{T,\varepsilon}$ means interaction effect of strain and deformation temperature with compensation coefficients.

The aim is to fit the function between deformation conditions and compensation coefficients. The compensation coefficients in different conditions are calculated and shown in Figure 9. It could be seen that the compensation coefficients and strain rate are positively correlated. The parameters n and $-(r_0 + r_1 T)$ in the F_{Voce} part are constant, but they are positively correlated with the strain rate by calculation. When the strain rate is increased, the value of the F_{Voce} part increases more slowly than the truth, resulting in a larger error at a higher strain rate. Moreover, as shown in Figure 9, the strain rate increases by 10 times, k increases by a certain multiple, which means the value of $K_{\dot{\varepsilon}}$ and logarithm of strain rate show an approximately linear relationship, and the $K_{\dot{\varepsilon}}$-value can be predicted by a logarithmic function. Meanwhile, a polynomial approach as follows was applied for describing the interaction among $k_{T,\varepsilon}$, strain, and temperature [44]:

$$k_{T,\varepsilon} = \left(A_0 + A_1 T + A_2 T^2 + \ldots + A_n T^n \right)\left(B_0 + B_1 \varepsilon + B_2 \varepsilon^2 + \ldots + B_n \varepsilon^n \right) \quad (16)$$

where n is the polynomial order. Fitting the functions in different orders used the correlation coefficient (R) [45] (shown in Equation (13)) to evaluate the fitting precision, and the results are shown in Table 2.

$$R = \frac{\sum_{i=1}^N (p_{Ei} - \overline{p_E})}{\sqrt{\sum_{i=1}^N (p_{Ei} - \overline{p_E})^2 \sum_{i=1}^N (\sum_{i=1}^N (p_{Pi} - \overline{p_P})^2}} \quad (17)$$

where p_{Ei} stands for the true value of K_c, p_{Pi} is the calculation by a polynomial function, $\overline{p_E}$ and $\overline{p_P}$ stand for the mean-value of the true value and calculated value, respectively.

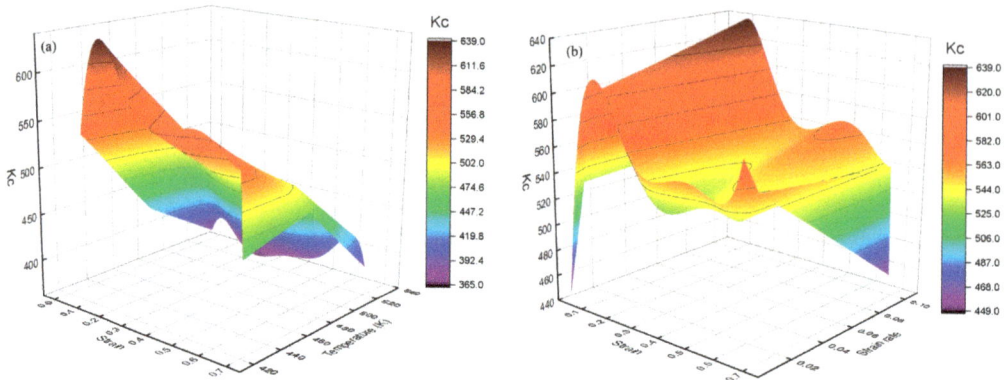

Figure 9. The value of Kc in different conditions (**a**) Kc in different strains and temperatures (**b**) Kc in different strains and strain rates.

Table 2. The correlation coefficient in different orders.

Polynomial Order	2	3	4	5
R	0.97787	0.99201	0.995424	0.99999

The correlation coefficients in the second to fifth orders are all above 0.95. To avoid the error of overfitting, a second-order polynomial function may be best for describing the interaction during the medium temperature range. The second-order polynomial function is shown in Figure 10b. Ultimately, the modified Voce-type constitutive model for 7050 alloy in medium-temperature compression is mathematically described below.

$$\sigma = K_c \, F_{Voce} \text{ MPa}$$

$$F_{Voce} = (\varepsilon - 0.02593)^{0.211942} \exp\left(\frac{142.13587}{T}\right) \exp(-(0.43411 + 0.000162213)\varepsilon) \, \dot{\varepsilon}^{(0.00683 + 0.0000183178T)}$$

$$K_C = \left(0.0275 \ln \dot{\varepsilon} + 1.0629\right)\left(-0.01252\varepsilon^2 + 0.01189\varepsilon - 0.02068\right)\left(0.4515T^2 - 321.3442T + 23582.594\right)$$

(18)

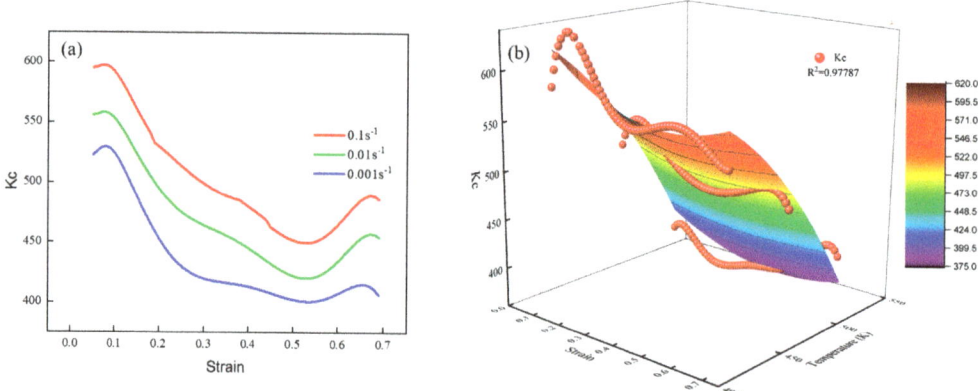

Figure 10. The relationship between (**a**) strain, strain rate, and Kc; (**b**) strain, temperature, and Kc.

3.3. Constitutive Model Evaluation and Application

A good constitutive model cannot only fit the modeling data used but also accurately calculate all the flow stresses during the actual compressive process. Therefore, the data used for the model establishment and the load of thermal compression were calculated and compared. The comparison between the predicted stresses by the conventional and corrected Voce models and the experimental values is shown in Figure 11a,b. The comparison indicates that the predicted stresses by the conventional Voce model have remarkable deviations from the measurements. However, the calculations by the modified model are well matched to the measured ones.

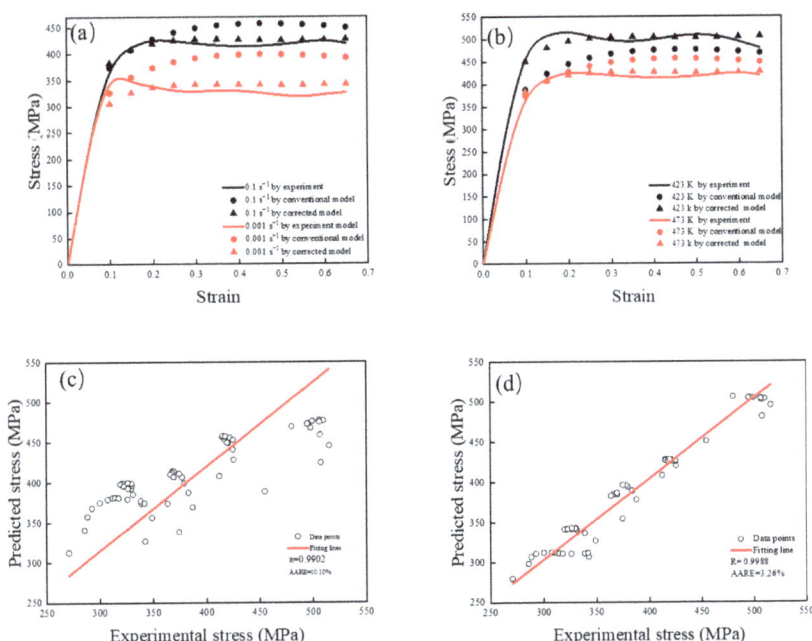

Figure 11. Comparisons between experimental and predicted stress by conventional and corrected Voce models: (**a**) 473 K at different strain rates; (**b**) 0.1 s^{-1} at different temperatures; (**c**) conventional model; (**d**) modified Voce model.

To compare the accuracy of the two models further, two statistical indexes, the correlation coefficient (R) and the average absolute relative error (AARE), were used as the evaluation criteria in this research [33]. The two indexes are expressed as follows:

$$\begin{cases} R = \dfrac{\sum_{i=1}^{N}(\sigma_{Ei}-\overline{\sigma_E})}{\sqrt{\sum_{i=1}^{N}(\sigma_{Ei}-\overline{\sigma_E})^2 \sum_{i=1}^{N}(\sum_{i=1}^{N}(\sigma_{Pi}-\overline{\sigma_P})^2}} \\ \text{AARE} = \dfrac{1}{N}\sum_{i=1}^{N}\left|\dfrac{\sigma_{Ei}-\sigma_{Pi}}{\sigma_E}\right| \times 100\% \end{cases} \quad (19)$$

where σ_{Ei} stands for the measurement, σ_{Pi} is the calculation by the constitutive model, and σ_E and σ_P stand for the mean value of the flow stress obtained by experiment and calculation, respectively.

Figure 11c,d show the comparison results for the two models. The R-value of the conventional Voce model and the AARE-value of the conventional Voce model are 0.9902 and 10.10%, respectively, while they are 0.9988 and 3.26%, respectively, for the modified model. Hence, the modified Voce model has better fitting ability than the convention for experimental data in a wide range.

The characterization of the flow behaviors of the 7050 aluminum alloy in its solid solution state contributes significantly to its warm forge process design. The final purpose of the modified Voce-type constitutive model is for better practical load prediction. In order to verify the prediction abilities of the practical loads of the modified model, the whole warm compression process was calculated and compared. The accuracy of flow stress correction and prediction for non-data points can also be evaluated by the FE simulation. The calculation was done on the Deform-2D 11.0 platform (Scientific Forming Technologies Corporation, Raleigh, North Carolina, USA). Here, we enrich the stress-strain data of the alloy utilizing the modified model. Then we simulated two compression processes at 333 K at different strain rates. The simulation model is shown in Figure 12. Only a representative plane was used for cylindrical compression. The size of the specimens is $\Phi 8$ mm $\times$ 12 mm. The aluminum specimen was assumed to be rigid viscoplastic and meshed with about 32,000 tetrahedral elements. The grid size was calculated by the simulation software from the curvature of the initial geometry. The deformation solver uses a conjugate gradient solver. It uses an iterative method to gradually approximate the optimal value. The parts were re-meshed between every 5 computational steps to avoid simulation errors since the mesh geometry changes significantly during compression. The heat transfer between the specimen and the air was neglected, and interface friction was characterized using the shear friction law with a friction factor of 0.2. Speed of top anvil and the strain rate of specimen can be converted as Equation (14) [46]:

$$V_T = \frac{H_B - H_A}{\varepsilon_T / \dot{\varepsilon}} \qquad (20)$$

where H_B is the height of the specimen before compression, and in this simulation, H_B is 12. H_A is the height after compression, namely 6 mm. ε_T is the true strain after compression, and it is approximately 0.69. $\dot{\varepsilon}$ is the strain rate of the specimen.

Figure 12. Finite element model for isothermal comparisons.

The simulation results are shown in Figure 13a,b. Due to friction and heterogeneous deformation, the periphery of the specimen became typical drum-type. At the end of this compression, the inner region indicates a true strain of 0.69, and the true strain rates are about 0.1 s^{-1} and 0.01 s^{-1}, respectively, manifesting the simulated physical quantities having good agreement with the compression conditions. The upsetting loads, along with the top die stroke, were calculated and compared. The load trend is basically consistent with the experimental result. According to Equation (13), the AARE of the two is 5.71% and 4.61%, respectively. Consequently, it shows the positive accuracy of the friction correction, and the modified Voce model can be effectively applied in the FEM for stress calculation and load prediction during warm forming of the solid solution-state 7050 aluminum alloy.

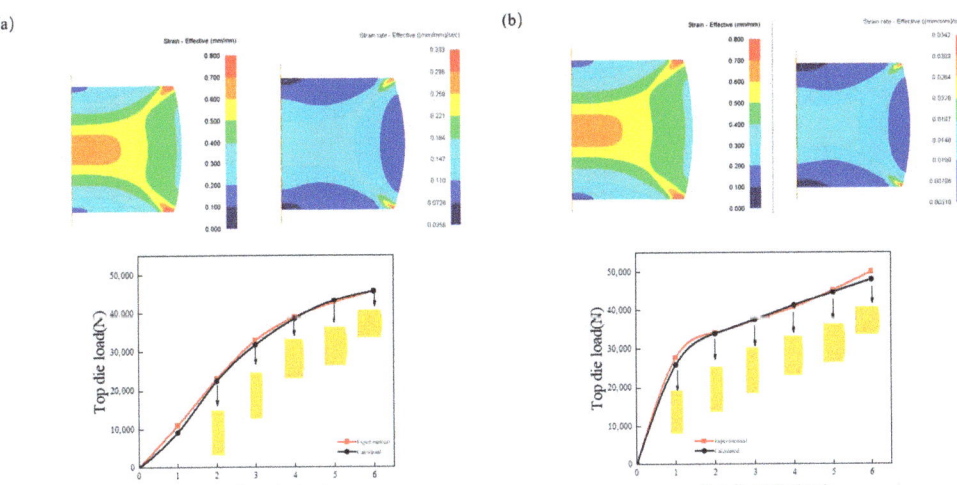

Figure 13. Comparison between simulated and measured load-stroke curves: (**a**) strain rate is $0.01\ \text{s}^{-1}$; (**b**) strain rate is $0.1\ \text{s}^{-1}$.

4. Conclusions

In this study, the room and warm compression behaviors of the solid solution-state 7050 alloy were researched. Furthermore, a modified Voce-type constitutive model with variable parameters was built to enhance the prediction precision of the warm compression behavior of solid solution-state 7050 alloy. Meanwhile, the prediction accuracy of the modified Voce is evaluated in two ways. The following conclusions could be obtained:

(1) The Voce-type model has better precision of warm compression behavior for solid solution-state 7050 alloy than the unsaturated models;

(2) The prediction accuracy of the conventional Voce model with softening coefficient on solid solution-state alloys during the warm compression process is not very high. In the conventional Voce model, the material parameters are regarded as independent of the deformation parameters. However, the parameters are variable due to microstructure deformation mechanism evolution, which decreases simulation accuracy;

(3) The modified model has good predictability for warm compression behavior for solid solution-state 7050 alloy. The fitting and FEA results indicate the validity of a modified Voce-type constitutive model with variable parameters and friction correction in the warm compression process. The modified Voce-type constitutive model can be applied to the warm forge process design of the 7050 alloy in its solid solution state.

Author Contributions: Investigation, methodology, data curation, and writing—original draft, H.T. and Y.X.; resources, conceptualization, formal analysis, and writing—review and editing, C.P. and Y.L. All authors have read and agreed to the published version of the manuscript.

Funding: The research was funded the National Natural Science Foundation of China (No. 51775068) and the Chongqing Natural Science Foundation general project [No. cstc2021jcyj-msxmX1085].

Institutional Review Board Statement: Not applicable.

Informed Consent Statement: Not applicable.

Data Availability Statement: The data presented in this study are not publicly available due to ongoing research in this field.

Conflicts of Interest: The authors declare that they have no known competing financial interest or personal relationships that could have appeared to influence the work reported in this paper.

References

1. Feng, D.; Xu, R.; Li, J.; Huang, W.; Wang, J.; Liu, Y.; Zhao, L.; Li, C.; Zhang, H. Microstructure Evolution Behavior of Spray-Deposited 7055 Aluminum Alloy During Hot Deformation. *Metals* **2022**, *11*, 1982. [CrossRef]
2. Wei, S.L.; Feng, Y.; Zhang, H.; Xu, C.T.; Wu, Y. Influence of Aging on Microstructure, Mechanical Properties and Stress Corrosion Cracking of 7136 Aluminum Alloy. *J. Cent. South Univ.* **2021**, *28*, 2687–2700. [CrossRef]
3. Wu, C.H.; Feng, D.; Ren, J.J.; Zang, Q.H.; Li, J.C.; Liu, S.D.; Zhang, X.M. Effect of Non-Isothermal Retrogression and Re-Ageing On through-Thickness Homogeneity of Microstructure and Properties of Al-8Zn-2Mg-2Cu Alloy Thick Plate. *J. Cent. South Univ.* **2022**, *29*, 960–972. [CrossRef]
4. Liu, Y.; Zhang, T.; Gong, H.; Wu, Y. Effect of Cold Pressing and Aging on Reduction and Evolution of Quenched Residual Stress for Al-Zn-Mg-Cu T-Type Rib. *Appl. Sci.* **2021**, *11*, 5439. [CrossRef]
5. Zhang, Z.; Yang, Y.; Li, L.; Chen, B.; Tian, H. Assessment of Residual Stress of 7050-T7452 Aluminum Alloy Forging Using the Contour Method. *Mater. Sci. Eng. A* **2015**, *644*, 61–68. [CrossRef]
6. Xu, L.; Zhan, L.; Xu, Y.; Liu, C.; Huang, M. Thermomechanical Pretreatment of Al-Zn-Mg-Cu Alloy to Improve Formability and Performance During Creep-Age Forming. *J. Mater. Process. Technol.* **2021**, *293*, 117089. [CrossRef]
7. Koc, M.; Culp, J.; Altan, T. Prediction of Residual Stresses in Quenched Aluminum Blocks and their Reduction through Cold Working Processes. *J. Mater. Process. Tech.* **2006**, *174*, 342–354. [CrossRef]
8. Rometsch, P.A.; Zhang, Y.; Knight, S. Heat Treatment of 7Xxx Series Aluminium Alloys—Some Recent Developments. *Trans. Nonferrous Met. Soc. China* **2014**, *24*, 15. [CrossRef]
9. Zhang, C.; Chu, X.; Guines, D.; Leotoing, L.; Ding, J.; Zhao, G. Dedicated linear–Voce model and its application in Dedicated Linear-Voce Model and its Application in Investigating Temperature and Strain Rate Effects on Sheet Formability of Aluminum Alloys. *Mater. Des.* **2014**, *67*, 522–530. [CrossRef]
10. Hartloper, A.R.; de Castro e Sousa, A.; Lignos, D.G. Constitutive Modeling of Structural Steels: Nonlinear Isotropic/Kinematic Hardening Material Model and its Calibration. *J. Struct. Eng.* **2021**, *147*. [CrossRef]
11. Sung, J.H.; Kim, J.H.; Wagoner, R.H. A Plastic Constitutive Equation Incorporating Strain, Strain-Rate, and Temperature. *Int. J. Plast.* **2010**, *26*, 1746–1771. [CrossRef]
12. Wu-Jiao, L.F.N.T. Effect of Solution Treatment Time on Mechanical Properties and Corrosion Resistance of Extruded 7050 Aluminum Alloy. *Trans. Mater. Heat Treat.* **2021**, *42*, 10.
13. Li, P.Y.; Xiong, B.Q.; Zhang, Y.A.; Li, Z.H.; Zhu, B.H.; Wang, F.; Liu, H.W. Quench Sensitivity and Microstructure Character of High Strength Aa7050. *Trans. Nonferrous Met. Soc. China* **2012**, *22*, 268–274. [CrossRef]
14. Feng, D.; Wang, G.; Chen, H.; Zhang, X. Effect of Grain Size Inhomogeneity of Ingot on Dynamic Softening Behavior and Processing Map of Al-8Zn-2Mg-2Cu Alloy. *Met. Mater. Int.* **2016**, *24*, 195–204. [CrossRef]
15. Robson, J.D.; Prangnell, P.B. Predicting Recrystallised Volume Fraction in Aluminium Alloy 7050 Hot Rolled Plate. *Met. Sci. J.* **2002**, *18*, 607–614. [CrossRef]
16. Fan, X. Homogenization of 7050 Plates by a Novel Differential Temperature Rolling. *Mater. Manuf. Process.* **2018**, *33*, 1822–1829. [CrossRef]
17. Han, N.M.; Zhang, X.M.; Liu, S.D.; He, D.G.; Zhang, R. Effect of Solution Treatment on the Strength and Fracture Toughness of Aluminum Alloy 7050. *J. Alloys Compd.* **2011**, *509*, 4138–4145. [CrossRef]
18. Juan, Y.; Niu, G.; Jiang, H.; Yang, J.; Tang, W.; Dai, Y.; Yang, Y.; Zhang, J. Influence of Solution Heat Treatment on Strength and Fracture Toughness of Aluminum Alloy 7050. *J. Cent. South Univ.* **2012**, *43*, 855–863.
19. Li, C.; Wang, H.; Wang, W.; Ye, C.; Jin, Z.; Zhang, X.; Zhang, H.; Jia, L. Effect of Solution Treatment on Recrystallization, Texture and Mechanical Properties of 7a65-T74 Aluminum Alloy Super-Thick Hot Rolled Plate. *J. Wuhan Univ. Technol.-Mater. Sci. Ed.* **2022**, *37*, 460–469. [CrossRef]
20. Zhou, Y.; Zhou, J.; Xiao, X.; Li, S.; Cui, M.; Zhang, P.; Long, S.; Zhang, J. Microstructural Evolution and Hardness Responses of 7050 Al Alloy During Processing. *Materials* **2022**, *15*, 5629. [CrossRef]
21. Amit, B. *Mechanical Properties and Working of Metals and Alloys*; Springer: Singapore, 2018.
22. Chen, S.; Chen, K.; Peng, G.; Chen, X.; Ceng, Q. Effect of Heat Treatment on Hot Deformation Behavior and Microstructure Evolution of 7085 Aluminum Alloy. *J. Alloys Compd.* **2012**, *537*, 338–345. [CrossRef]
23. Gasson, C.P. Light Alloys: From Traditional Alloys to Nanocrystals—Fourth Edition. *Aeronaut. J.* **2006**, *110*, 394–395. [CrossRef]
24. Lu, J.; Song, Y.; Hua, L.; Zheng, K.; Dai, D. Thermal Deformation Behavior and Processing Maps of 7075 Aluminum Alloy Sheet Based on Isothermal Uniaxial Tensile Tests. *J. Alloys Compd.* **2018**, *767*, 856–869. [CrossRef]
25. Zhu, R.H.; Qing LI, U.; Li, J.F.; Chen, Y.L.; Zhang, X.H.; Zheng, Z.Q. Flow Curve Correction and Processing Map of 2050 Al-Li Alloy. *Trans. Nonferrous Met. Soc. China* **2018**, *28*, 404–414. [CrossRef]
26. Luo, J.; Li, M.Q.; Ma, D.W. The Deformation Behavior and Processing Maps in the Isothermal Compression of 7a09 Aluminum Alloy. *Mater. Sci. Eng. A* **2012**, *532*, 548–557. [CrossRef]
27. Zhao, J.; Deng, Y.; Tang, J.; Zhang, J. Influence of Strain Rate on Hot Deformation Behavior and Recrystallization Behavior Under Isothermal Compression of Al-Zn-Mg-Cu Alloy. *J. Alloys Compd.* **2019**, *809*, 151788. [CrossRef]
28. Liu, L.; Wu, Y.X.; Gong, H.; Wang, K. Modification of Constitutive Model and Evolution of Activation Energy On 2219 Aluminum Alloy During Warm Deformation Process. *Trans. Nonferrous Met. Soc. China* **2019**, *29*, 448–459. [CrossRef]

29. Jiang, F.; Tang, J.; Fu, D.; Huang, J.; Zhang, H. A Correction to the Stress–Strain Curve During Multistage Hot Deformation of 7150 Aluminum Alloy Using Instantaneous Friction Factors. *J. Mater. Eng. Perform.* **2018**, *27*, 3083–3090. [CrossRef]
30. Wu, D.X.; Long, S.; Li, S.S.; Zhou, Y.T.; Wang, S.Y.; Dai, Q.W.; Lin, H.T. Hot Deformation Behavior of Homogenized Al-7.8Zn-1.65Mg-2.0Cu (Wt.%) Alloy. *J. Mater. Eng. Perform.* **2022**, *32*, 3431–3442. [CrossRef]
31. Hollomon, J.H. Tensile Deformation. *Met. Technol.* **1945**, *12*, 268–290.
32. Swift, H.W. Plastic Instability Under Plane Stress. *J. Mech. Phys. Solids* **1952**, *1*, 1–18. [CrossRef]
33. Ludwigson, D. Modified Stress-Strain Relation for Fcc Metals and Alloys. *Metall. Trans.* **1971**, *2*, 2825. [CrossRef]
34. Voce, E. The Relationship Between Stress and Strain for Homogeneous Deformation. *J. Inst. Met.* **1948**, *74*, 537–562.
35. Verlinden, B.; Suhadi, A.; Delaey, L. A Generalized Constitutive Equation for an Aa6060 Aluminium Alloy. *Scr. Metall. Mater.* **1993**, *28*, 1441–1446. [CrossRef]
36. Pelaccia, R.; Santangelo, P.E. A Homogeneous Flow Model for Nitrogen Cooling in the Aluminum-Alloy Extrusion Process. *Int. J. Heat Mass Transf.* **2022**, *195*, 1–14. [CrossRef]
37. Shi, S.X.; Liu, X.S.; Zhang, X.Y.; Zhou, K.C. Comparison of Flow Behaviors of Near Beta Ti-55511 Alloy During Hot Compression Based on Sca and Bpann Models. *Trans. Nonferrous Met. Soc. China* **2021**, *31*, 1665–1679. [CrossRef]
38. Nourani, M.; Milani, A.S.; Yannacopoulos, S. On the Effect of Different Material Constitutive Equations in Modeling Friction Stir Welding: A Review and Comparative Study on Aluminum 6061. *Int. J. Adv. Eng. Technol.* **2014**, *7*, 1–20.
39. Koc, P.; Atok, B. Computer-Aided Identification of the Yield Curve of a Sheet Metal After Onset of Necking. *Comput. Mater. Sci.* **2004**, *31*, 155–168. [CrossRef]
40. Ye, J.H.; Chen, M.H.; Wang, N.; Xie, L.S. Flow Stress Model of 2a12 Aluminum Alloy Based on Modified Voce Model. *Trans. Mater. Heat Treat.* **2019**, *40*, 170–176.
41. Wei, G.; Peng, X.; Hadadzadeh, A.; Mahmoodkhani, Y.; Xie, W.; Yang, Y.; Wells, M.A. Constitutive Modeling of Mg-9Li-3Al-2Sr-2Y at Elevated Temperatures. *Mech. Mater.* **2015**, *89*, 241–253. [CrossRef]
42. Jedrasiak, P.; Shercliff, H. Finite Element Analysis of Small-Scale Hot Compression Testing—Sciencedirect. *J. Mater. Sci. Technol.* **2021**, *76*, 174–188. [CrossRef]
43. Zhang, J.S.; Xia, Y.F.; Quan, G.Z.; Wang, X.; Zhou, J. Thermal and Microstructural Softening Behaviors During Dynamic Recrystallization in 3Cr20Ni10W2 Alloy. *J. Alloys Compd.* **2018**, *743*, 464–478. [CrossRef]
44. Padhi, P.C.; Mahapatra, S.S.; Yadav, S.N.; Tripathy, D.K. Multi-Objective Optimization of Wire Electrical Discharge Machining (Wedm) Process Parameters Using Weighted Sum Genetic Algorithm Approach. *J. Adv. Manuf. Syst.* **2016**, *15*, 85–100. [CrossRef]
45. Long, S.; Xia, Y.-F.; Wang, P.; Zhou, Y.-T.; Gong-Ye, F.-J.; Zhou, J.; Zhang, J.-S.; Cui, M.-L. Constitutive Modelling, Dynamic Globularization Behavior and Processing Map for Ti-6Cr-5Mo-5V-4Al Alloy During Hot Deformation. *J. Alloys Compd.* **2019**, *796*, 65–76. [CrossRef]
46. Wu, D.; Long, S.; Wang, S.; Li, S.S.; Zhou, Y.T. Constitutive Modelling with a Novel Two-Step Optimization for an Al-Zn-Mg-Cu Alloy and its Application in Fea. *Mater. Res. Express* **2021**, *8*, 116511. [CrossRef]

Disclaimer/Publisher's Note: The statements, opinions and data contained in all publications are solely those of the individual author(s) and contributor(s) and not of MDPI and/or the editor(s). MDPI and/or the editor(s) disclaim responsibility for any injury to people or property resulting from any ideas, methods, instructions or products referred to in the content.

Article

A New Method for Preparing Titanium Aluminium Alloy Powder

Jialong Kang [1,2], Yaoran Cui [1,2], Dapeng Zhong [1,2], Guibao Qiu [1,2,*] and Xuewei Lv [1,2]

[1] Chongqing Key Laboratory of Vanadium-Titanium Metallurgy and Advanced Materials, Chongqing University, Chongqing 400044, China; 17748351783@163.com (J.K.); 18883158379@163.com (Y.C.); zhongdapengcqu@163.com (D.Z.); lvxuewei@163.com (X.L.)
[2] College of Materials Science and Engineering, Chongqing University, Chongqing 400044, China
* Correspondence: qiuguibao@cqu.edu.cn

Abstract: Due to TiAl alloys' excellent properties, TiAl alloys have received widespread attention from researchers. However, the high energy consumption and lengthy process of traditional preparation methods have always limited the large-scale application of TiAl alloys. This article develops a new method for preparing TiAl-based alloy powder via the magnesium thermal reduction of TiO_2 in $AlCl_3$-KCl molten salt. In this study, the proportion of $AlCl_3$&KCl molten salts was determined. We conducted phase analysis on the final product by studying the changes in temperature and time. It was found that the $TiAl_3$ alloy powder could be obtained by being kept at 750 °C for 2 h, with an oxygen content of 3.91 wt%. The reaction process for the entire experiment was determined through thermodynamic calculations and experimental analysis, and the principles of the reduction process are discussed.

Keywords: TiAl alloy; magnesium reduction; $AlCl_3$-KCl; TiO_2

Citation: Kang, J.; Cui, Y.; Zhong, D.; Qiu, G.; Lv, X. A New Method for Preparing Titanium Aluminium Alloy Powder. *Metals* **2023**, *13*, 1436. https://doi.org/10.3390/met13081436

Academic Editor: Wislei Riuper Osório

Received: 5 July 2023
Revised: 2 August 2023
Accepted: 7 August 2023
Published: 10 August 2023

Copyright: © 2023 by the authors. Licensee MDPI, Basel, Switzerland. This article is an open access article distributed under the terms and conditions of the Creative Commons Attribution (CC BY) license (https://creativecommons.org/licenses/by/4.0/).

1. Introduction

Metal compounds offer both the plasticity of metals, and the high-temperature strength of ceramics in specific compositions, due to the metal compound ordered arrangement of their atoms, and the coexistence of inter-atomic metal bonds and covalent bonds. TiAl-based intermetallic compounds were highly valued in the aerospace industry and other fields, due to their excellent mechanical properties and low density, in the 1950s [1,2]. Nowadays, the TiAl-based alloy is still recognized as a high-end material in the world [3,4]. Due to its high specific strength, high temperature resistance [5,6], corrosion resistance, oxidation resistance [7,8], and excellent biocompatibility, the TiAl-based alloy is widely used in the field of high-end materials in modern life [9,10]. For example, large aeroplanes, submarines, aerospace technology, and artificial bones [11–13]. This high-end material should also be popularized in ordinary daily life but, due to its high cost, it has not been applied, and can only be applied in high- and middle-value fields. The reason for this dilemma is the long cycle and high cost of preparing the metal Ti. Since the discovery of metallic titanium, only the Kroll method has produced sponge titanium on a large scale [14–17]. Therefore, most methods for preparing TiAl-based alloys are element approaches [18–21], prepared by adding proportional amounts of the elements Ti and Al in a high-temperature melting furnace. Adding Ti separately also causes production costs for TiAl-based alloys.

Many researchers have adopted different methods to find an efficient method for producing TiAl-based alloys. The most traditional way is to prepare TiAl-based alloys via casting [22–25] and cast alloys with different compositions using a vacuum induction melting furnace, centrifuge, hot press, and other equipment. Among them, J Lapin [26] et al. prepared the Ti-42.6Al-8.7Nb-0.3Ta-2.0C and Ti-41.0Al-8.7Nb-0.3Ta-3.6C (in at.%) TiAl alloys via the casting method, and studied the effect of adding 2.0C and 3.6C on the properties of the TiAl alloys. They also studied the solid-state phase transformation grain refinement of the as-cast peritectic TiAl-based alloys. Powder metallurgy is also the preparation method for most TiAl-based alloys; it can overcome the defects generated through traditional manufacturing methods, and obtain uniform and fine microstructures, significantly improving the mechanical properties of the alloys. Heike Gabrisch et al. [27] added 0.5–1.0 at.% C into Ti-45Al-5Nb alloy via powder metallurgy, and used a transmission electron microscope and high-energy XRD to study the influence of solid-solution carbon and carbide precipitation on the hardness of the TiAl alloy. In addition to the above two methods, emerging additive manufacturing technologies [28–30] can also prepare designed alloys, by stacking powders layer by layer. The powder preparation methods mentioned above all require titanium powder as a raw material, which is expensive, and increases production costs. Therefore, there is a need for a method that can directly prepare alloys from TiO_2, to improve the existing technology, reduce the process flow, and increase the popularity of TiAl-based alloys [31,32].

Researchers have now prepared TiAl-based alloys using other methods. Zhao et al. [33,34] proposed a two-step thermal reduction method for preparing TiAl-based alloys. In this method, firstly, Na_2TiF_6 is reduced by TiAl-based alloy powder, which is the first reduction stage. After vacuum distillation, the second-stage reduction of Al is carried out, to obtain TiAl-based alloys. The TiAl-based alloy powder collected in the purification process can also be used for the first-stage reduction. This method successfully realizes the overall round-robin preparation of TiAl-based alloys through the Al thermal reduction of Na_2TiF_6. Dou, Song, and Zhang [35,36] successfully prepared 20 kg TiAl-based alloy ingots with an oxygen content of approximately 1.09 wt% through multi-stage profound reduction. This method successfully achieved the one-step preparation of TiAl-based alloys through adding $KClO_3$ as a heating agent, and self-propagating aluminium heat.

However, the two-step aluminothermic reduction of Na_2TiF_6, and the high-temperature self-propagation method, require experiments at high temperatures, resulting in energy consumption. However, due to the limited reduction effect of metal aluminium on TiO_2 at low temperatures, it is impossible to prepare TiAl-based alloys by reducing TiO_2 at low temperatures. Moreover, TiAl-based alloy ingots exhibit significant room temperature brittleness, with a derivation rate of less than 1%, and fracture during stretching, resulting in a poor machinability, and difficulties in their application. Due to the significant difference in melting points between Ti and Al, casting high-quality alloy ingots is difficult and costly. This article proposes a new method of preparing TiAl-based alloy powder by reducing TiO_2 in $AlCl_3$&KCl molten salt through a magnesiothermic reduction. This method can only be carried out at low temperatures, with a simple production method, and a low level of environmental pollution. This work provides new ideas for the future large-scale application of TiAl-based alloy powder, and to solve the problem of the long preparation process of existing TiAl-based alloys.

2. Materials and Method

2.1. Materials

Anhydrous $AlCl_3$, KCl, HCl, TiO_2, and binder were obtained from Aladdin Reagent Co., Ltd. (Shanghai, China). The particle size of anhydrous $AlCl_3$, KCl, and TiO_2 is below 74 μm. The binder used in this paper is ethyl cellulose, a polymer compound with the chemical formula $(C_{12}H_{22}O_5)_n$. Ethyl cellulose is a white powder at room temperature. The primary function of this binder is to aggregate the raw materials, and increase the contact area of raw materials. Under high-temperature conditions, ethyl cellulose will be decomposed into organic compounds that do not impact the experiment. As a reducing agent, Mg powder is obtained from Sinopharm Chemical Reagent Co., Ltd. (Ningbo, China). The Mg powder particle size is below 200 μm.

2.2. Introduction to AlCl₃ and Selection of AlCl₃-KCl Molten Salt Ratio

AlCl₃ is a white crystalline powder with a solid hydrochloric acid odour, and a light yellow industrial product, and is easily soluble in water, alcohol, chloroform, etc. Its melting point is 194 °C, but it is prone to sublimation and deliquescence at 178 °C. Moreover, due to the exothermic hydration reaction, it may explode when encountering water. Therefore, the AlCl₃ must be sealed and stored in a dry environment.

(1) Physical properties of AlCl₃

AlCl₃ was first prepared by Biltz. W [37] in 1923, via the reaction of 99.5% pure Al powder with dry HCl gas. Subsequently, Baker Analytics and Grothe studied the basic properties of AlCl₃, etc. [37]. The density and viscosity of AlCl₃ are key physical properties during the thermal reduction process. Based on previous research data, the density of AlCl₃ varies with the temperature, as shown in Figure 1. It can be seen that the density of AlCl₃ shows a steady downward trend with the temperature rising within 460–560 °C. The following formula is the density formula, fitted according to the data (estimated standard error: 0.15%).

$$\rho = 2.19 - 5.40 \times 10^{-4}T - 1.14 \times 10^{-5}T^2 + 3.04 \times 10^{-8}T^3 \tag{1}$$

ρ indicates the density of AlCl₃ (g·cm⁻³); T indicates the temperature (°C).

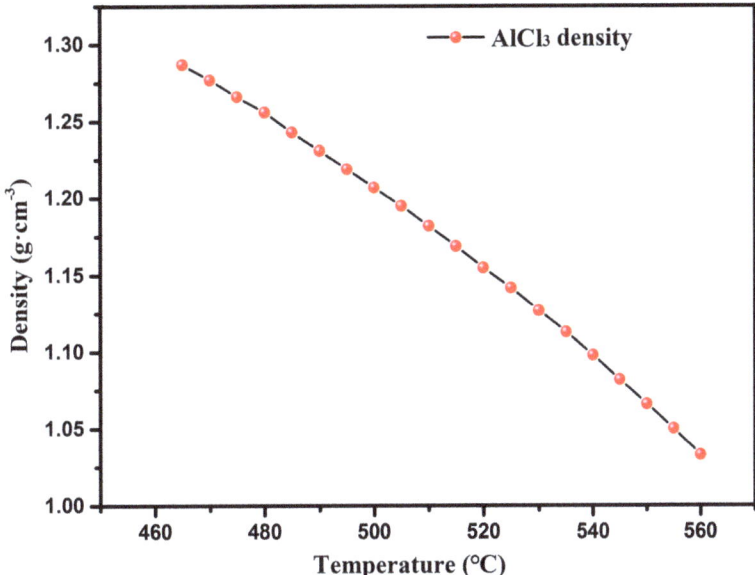

Figure 1. AlCl₃ density changes with temperature, data from [37].

Figure 2 shows the viscosity data of AlCl₃. According to these data, the viscosity formula of AlCl₃ is fitted, and the estimated standard error is 1.05%.

$$\eta = 1.71 \cdot \exp(4943.8/RT) \tag{2}$$

η indicates the viscosity of AlCl₃ (Pa·s), and R indicates a constant of 8.314 (kJ/mol); T indicates the temperature (°C).

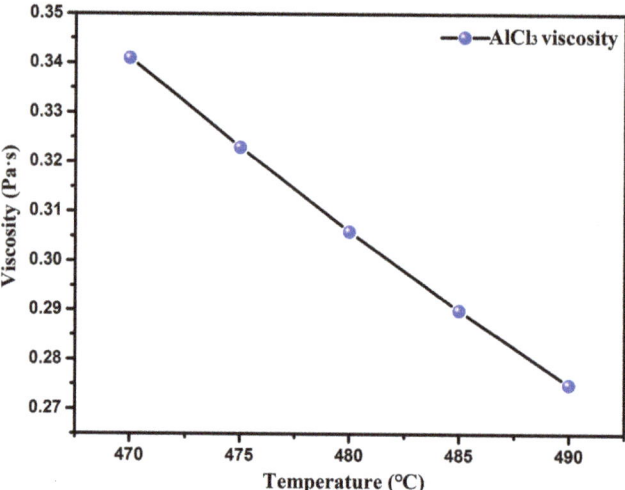

Figure 2. AlCl$_3$ viscosity changes with temperature, data from [37].

From the graph showing the viscosity and density changes with temperature, it can be seen that the density and viscosity of AlCl$_3$ show an overall decreasing trend with an increasing temperature. The decrease in density and viscosity is conducive to the complete contact of reactants during the reduction reaction process, increasing the reaction efficiency, which is a favourable factor in magnesium thermal reduction.

(2) Basic Physical Properties of AlCl$_3$-KCl Mixed Molten Salt

Due to the low melting point and easy sublimation of AlCl$_3$, to prevent the loss of raw materials due to the large amount of sublimation and volatilization of AlCl$_3$, a mixed molten salt AlCl$_3$-KCl is configured, which can effectively suppress the volatilization of AlCl$_3$. As shown in Figure 3, when the AlCl$_3$ content accounts for 80 wt% and above, the binary phase diagram of AlCl$_3$-KCl shows that a large amount of AlCl$_3$ gas is generated. As the KCl increases, the blue part in the figure shows the liquid phase zone of the AlCl$_3$-KCl eutectic salt without gas generation. Therefore, this part can be selected as the area used in the raw material ratio in this experiment, where the mass ratio of AlCl$_3$/(AlCl$_3$-KCl) is 0.65.

The density of the AlCl$_3$-KCl is based on Carter and Morrey's [37] work to obtain the following AlCl$_3$-KCl density data, as shown in Table 1. The temperature dependence of AlCl$_3$-KCl under different KCl contents was plotted using Table 1, as shown in Figure 4. It can be seen that as the KCl content increases, the overall density of the AlCl$_3$-KCl shows an upward trend. As the temperature increases, the density of the AlCl$_3$-KCl decreases. When the molar ratio of KCl exceeds 50%, the overall density of the AlCl$_3$-KCl does not change significantly with temperature. The incremented density is due to the melting point of KCl being higher. After the KCl content increases, the overall melting point of the AlCl$_3$-KCl increases. An increasing KCl content can inhibit AlCl$_3$ volatilization. This result is consistent with the calculation results in Figure 4.

Table 1. The density of the AlCl$_3$-KCl varies with the KCl content.

Mol % KCl	$\rho = a + bT$		Standard Error
	a	b·10^{-3}	
20.00	2.0252	−1.0038	0.11%
33.33	1.9889	−0.7901	0.14%
50.03	1.9556	−0.6622	0.05%
66.66	1.9734	−0.6101	0.02%

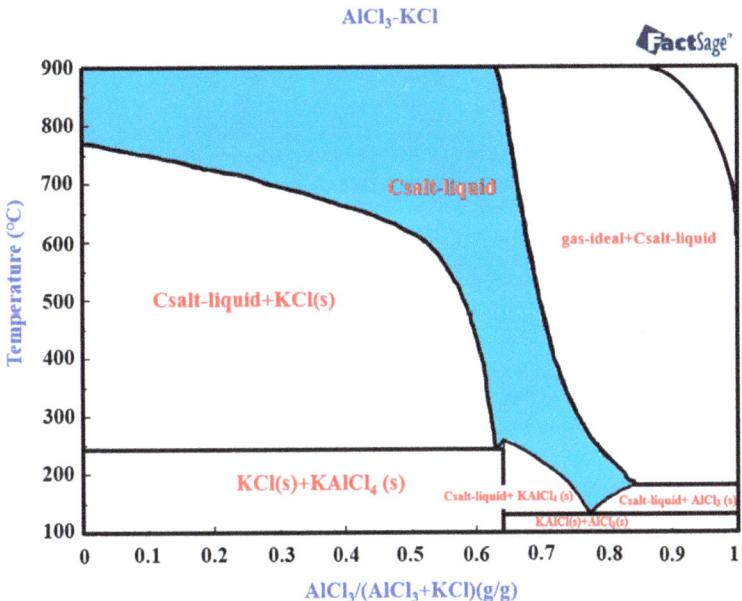

Figure 3. Binary phase diagram of $AlCl_3$-KCl.

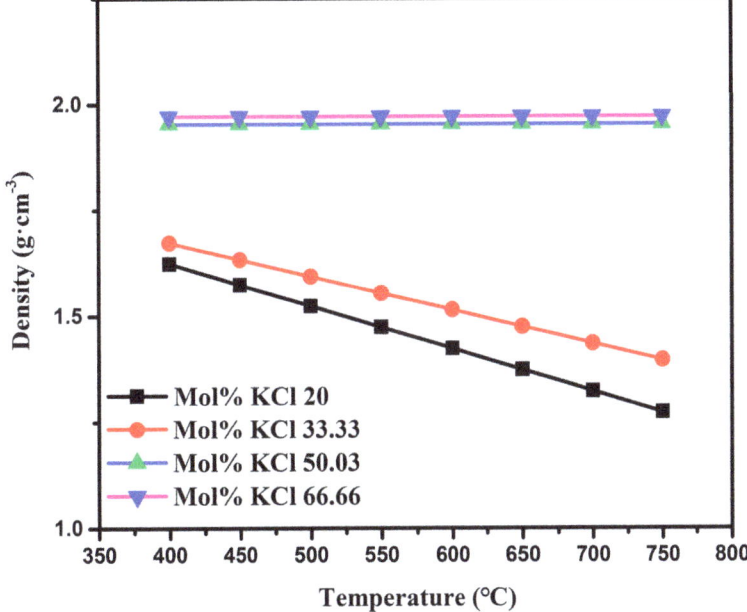

Figure 4. Density variation in $AlCl_3$-KCl with the temperature, under different KCl contents.

2.3. Methods

The raw materials are mixed with the binder in a mortar. After that, the mixed raw materials are pressed into a round green mass. For the reduction, we place the green mass into the molybdenum crucible and the tube furnace (KF1100 Nanjing Boyuntong (Nanjing, China), as shown in Figure 5). After reaching the specified temperature, this furnace can be loaded into the reaction furnace tube. The heating rate is 15 °C/min, and the insulation is maintained after reaching the specified temperature. According to the analysis of the physical properties and phase diagram of $AlCl_3$&KCl in the previous section, to suppress the volatilization of $AlCl_3$, the molten salt is weighed based on the mass ratio of $AlCl_3/(AlCl_3+KCl)$ of 0.65. Therefore, this experiment's molten salt mass ratio of $AlCl_3$:KCl is 1.8. Considering that $AlCl_3$ has the characteristic of volatilizing at low temperatures, it is necessary to increase the content of $AlCl_3$ appropriately during the raw material configuration process. The amount of TiO_2 used in each experiment is 5 g, the proportion of molten salt is four times the total mass of the TiO_2 added, and the amount of magnesium added is the molar mass of the $AlCl_3$ completely reacted.

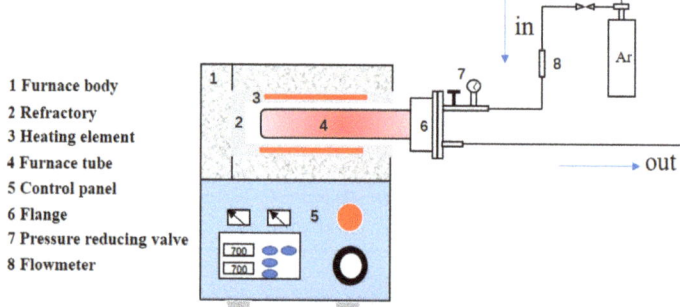

Figure 5. Experimental device.

The experimental process is shown in Figure 6. The experiment was conducted in a tubular box mixing furnace. The raw materials were mixed evenly, loaded into a molybdenum crucible, and placed in a reactor. Inert gas was first introduced into the reactor to exhaust the air, with a flow rate of 300 mL/min. After the high-temperature furnace rose to the set temperature, the reactor was loaded into the furnace, without increasing the temperature of the furnace (to reduce the evaporation of the molten salt). This experiment explored the influence of time on the reaction products' phase and oxygen content. The overall reaction time starts from the installation of the reaction furnace tube into the furnace, and a specific insulation time is set. After the reaction, argon gas is continuously introduced and cooled in the furnace, to ensure that the sample does not come into contact with oxygen. The final reactant is extracted for subsequent acid leaching and flotation, to obtain the final alloy product for analysis. Among the experiments, 5 wt% dilute HCl is used for acid leaching. The acid-leaching process is carried out in a water bath at 80 °C for 2 h. The acid-leaching process is mainly used to remove Mg and MgO. The specific reactions are shown in the formulae below. The reaction byproduct Al_2O_3 is removed via ball milling. The mass ratio of the ball-milling process as ball:material:H_2O during ball milling is 1.6:1:3. The ball mill rotates at a speed of 350 r/min for 15 min each time. Each experiment requires ball milling at least three times.

$$Mg + 2HCl = MgCl_2 + H_2 \qquad (3)$$

$$MgO + 2HCl = MgCl_2 + H_2O \qquad (4)$$

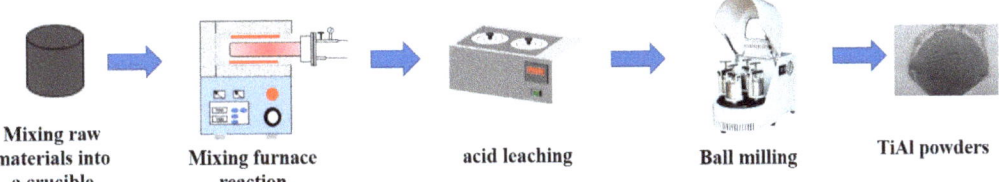

Figure 6. Experimental flowchart.

2.4. Analysis

X-ray powder diffraction (XRD, Cu Kα radiation, PANalytical X'Pert Powder, Malvern PANalytical B.V., Almelo, The Netherlands), with a scanning range of 10–90° and a scanning step of 5 deg./min, was carried out, to confirm the phase composition of sample powder three times. Scanning electron microscopy (SEM, TESCAN VEGA 3 LMH system, TESCAN, Brno, Czech Republic) was also performed, to evaluate the microanalysis and surface morphology of the reaction product. Energy-dispersive spectroscopy (EDS) microanalysis was conducted. The accelerating voltage used for the EDS analysis was 10 kV. The thermodynamic software FactSage8.0 calculated the feasibility of the reaction. The oxygen content of the product was analyzed using the JXA82 electron probe from JEOL Corporation, and the particle size analyzer from NanoBrook Omni of Brookhaven Instruments, Holtsville, NY, USA.

3. Results and Discussion

3.1. Phase Analysis of Reactants at Different Temperatures

The XRD phase diagrams and SEM of the products at different reaction temperatures are shown in Figures 7 and 8. Comparing the XRD analyses of the products at reaction temperatures of 750~950 °C, it can be found that the reaction products are TiAl-based alloy powders. Compared with traditional magnesiothermic reduction, TiO_2 is not directly reduced; other substances participate in the reaction. The Gibbs free energy diagram of the reaction between Mg and $AlCl_3$, TiO_2, was drawn using the thermodynamic software FactSage8.0, as shown in Figure 9. At temperatures ranging from 400 °C to 950 °C, the Gibbs free energy of the reaction between Mg and $AlCl_3$, TiO_2 can be observed. At the same magnesium content, Mg preferentially reacts with $AlCl_3$, and the entire process uses aluminium as a reducing agent in reducing the TiO_2. Moreover, magnesium-reducing $AlCl_3$ releases a large amount of heat, to drive the aluminothermic reduction of TiO_2. Therefore, the entire reaction process is as follows:

$$3Mg + 2AlCl_3 \rightarrow 2Al + 3MgCl_2 \quad (5)$$

$$Al + TiO_2 \rightarrow Ti + Al_2O_3 \quad (6)$$

$$6Ti + 6Al \rightarrow 4Ti + 2TiAl_3 \quad (7)$$

$$4Ti + 2TiAl_3 \rightarrow Ti_3Al + TiAl + 2TiAl_2 \quad (8)$$

$$Ti_3Al + 2TiAl_2 + TiAl \rightarrow 6TiAl \quad (9)$$

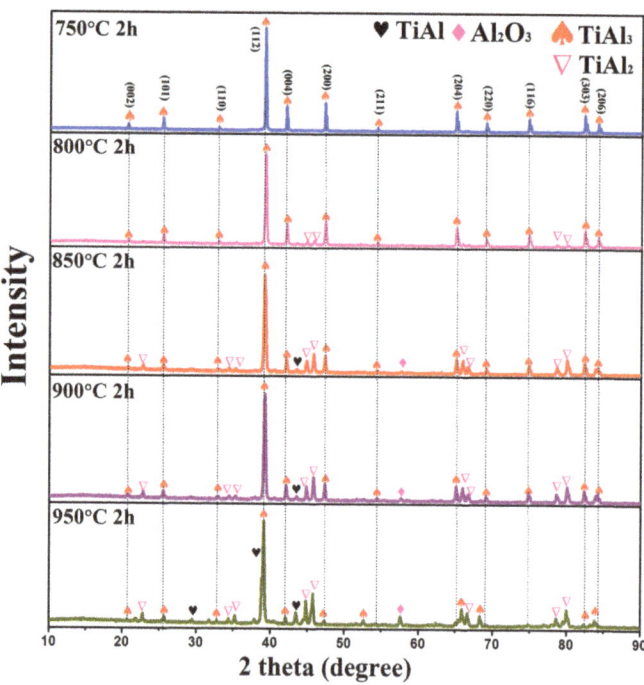

Figure 7. XRD patterns of products at different temperatures.

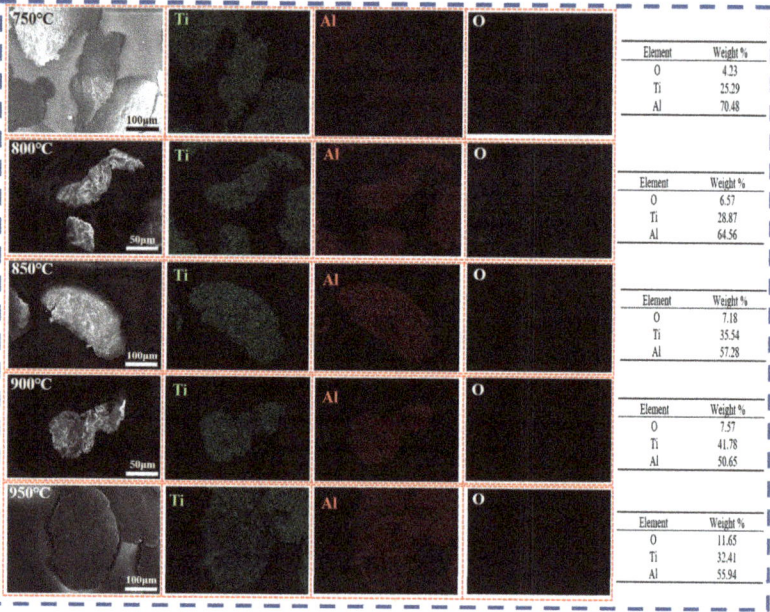

Figure 8. SEM and surface scanning of magnesium thermal reduction products, assisted by the $AlCl_3$&KCl molten salt medium.

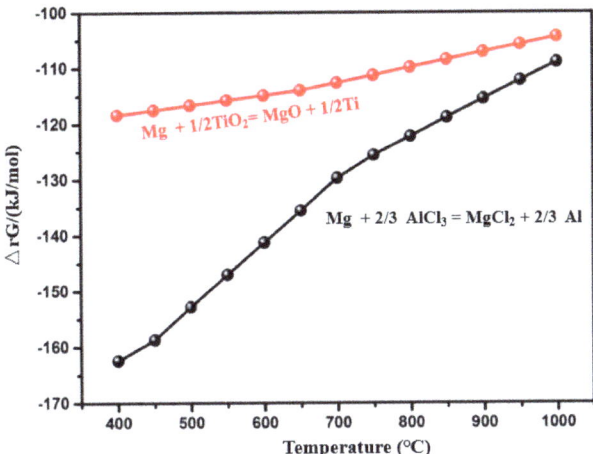

Figure 9. Gibbs free energy of the reaction between AlCl$_3$ and TiO$_2$, with Mg.

According to literature reports [38], the alloy reaction between Al and Ti at low temperatures is mainly liquid–solid. In this experiment, the main product phase is the TiAl$_3$ alloy after 2 h of reaction at 750 °C. As the reaction temperature increases, from 800 °C to 950 °C, the peak of TiAl$_3$ gradually decreases, while the peak of TiAl$_2$ gradually increases. The peak of TiAl$_2$ gradually increases because external Al reacts before Ti to generate TiAl$_3$ during the reaction process, and then gradually forms TiAl$_2$, as shown in Equation (8). The increase in temperature is more conducive to the formation of the TiAl alloy. From the energy spectrum detected via SEM surface scanning, it can be seen that the obtained TiAl alloy has a uniform distribution of components, with an oxygen content of 4.23 wt%. The reason for such a high oxygen content is that the reduction effect of Al is limited, and TiO$_2$ cannot be reduced to a lower oxygen content state. However, due to the presence of Mg, some O is absorbed, to some extent. As the reaction temperature increases, it is found that the oxygen content in the reaction product is low, at 750 °C.

3.2. Effect of Reduction Temperature

In the previous section, the study of different reaction temperatures in the magnesiothermic reduction of TiO$_2$ in the AlCl$_3$&KCl molten salt system found that the oxygen content of the TiAl$_3$-based alloys generated at 750 °C was relatively low and stable. Therefore, 750 °C was chosen as the optimal reaction temperature. This section takes the reaction time as a single variable, and experimentally studies the effect of different reaction times on the preparation of TiAl-based alloys at 750 °C. We set the reaction time at 750 °C for 1, 2, and 4 h, respectively, to study the effect of the reaction time on the reduction effect during the reaction process. Figures 10 and 11 show that XRD and SEM characterized the reaction products. After reacting at 750 °C for 1 h, the main phases of the reaction products are TiAl$_3$ and metallic Al. The TiAl$_3$ alloy is caused by the short reaction time between Al and Ti, and some of the reduced Al has not yet formed an alloy with Ti. Therefore, a large amount of metal Al has not reacted in the product after one hour of reaction. With the extension of the reaction time, when the reaction temperature is 2 h, the main product of the reaction is TiAl$_3$. After further prolonging the reaction time to 4 h, the main reaction products are the TiAl alloy, and a small amount of Al$_2$O$_3$. The occurrence of Al$_2$O$_3$ is because, with the extension of the reaction time, some Al forms Al$_2$O$_3$ powder through solid-state diffusion with the alloy, after reducing the TiO$_2$. This Al$_2$O$_3$ cannot be removed from the surface of the alloy through acid leaching and flotation. The attached Al$_2$O$_3$ can be seen in the SEM image in Figure 12; a small amount of matte and rough surface substance is attached to the metal surface, which is the Al$_2$O$_3$ generated through the extended reaction time.

As mentioned above, a stable TiAl$_3$ alloy powder can be obtained by reacting TiO$_2$ with AlCl$_3$&KCl molten magnesium salt for 2 h at 750 °C.

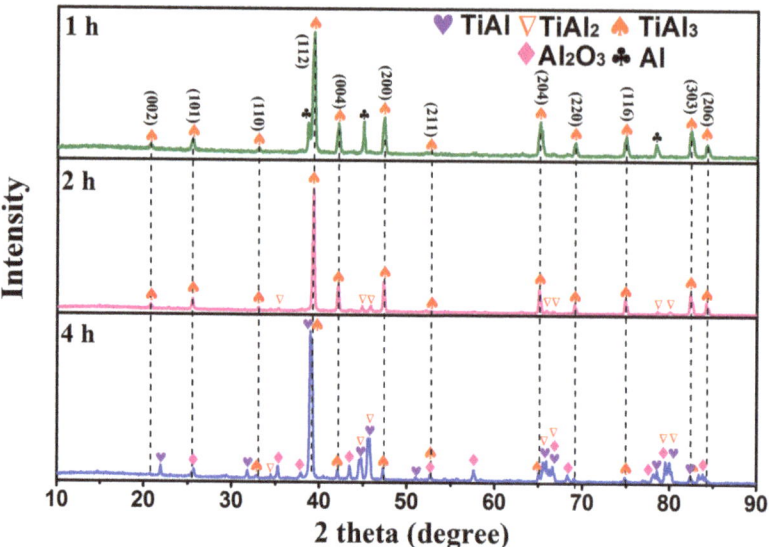

Figure 10. Effect of different reduction times on the reduction products.

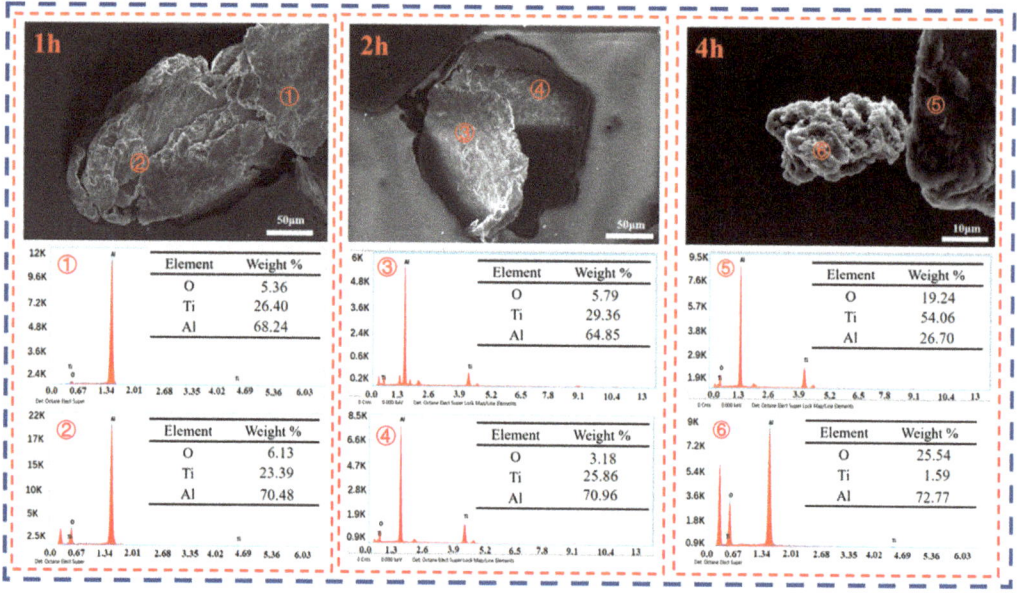

Figure 11. SEM and EDS of products with different reduction times.

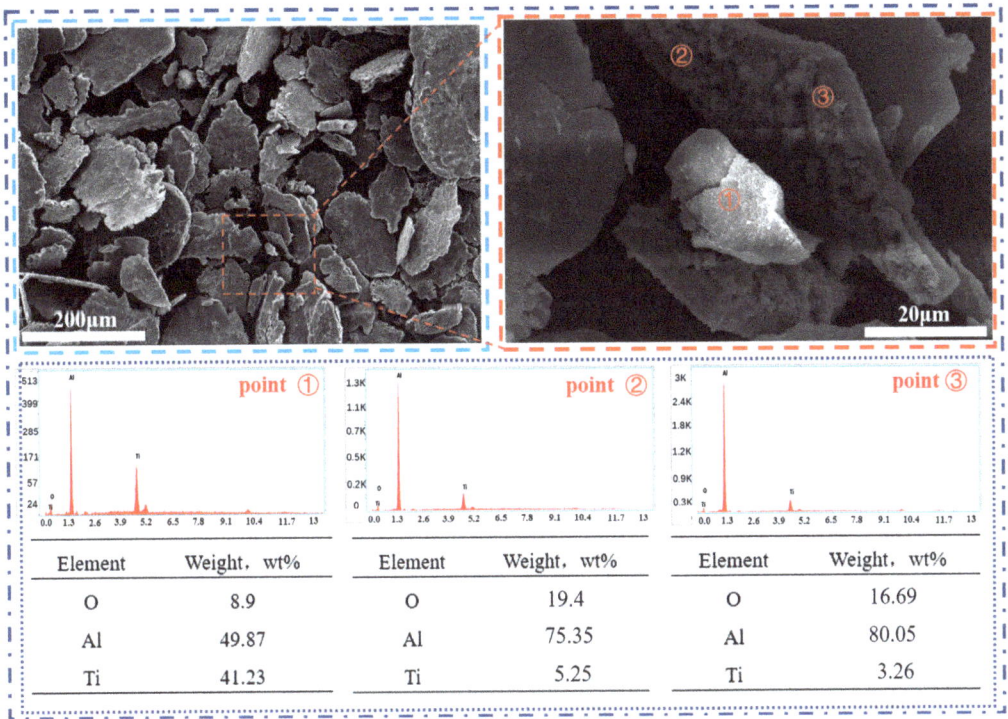

Figure 12. Scanning electron microscopy (SEM) images and energy-dispersive spectroscopy (EDS) of the products at 950 °C.

As the reaction time increases, the oxygen content in the reaction product gradually increases. At the same time, this is also because the solid-phase-to-solid-phase diffusion reaction between the Al_2O_3 and TiAl-based alloys occurs at high temperatures, which finally causes part of the Al_2O_3 to enter the TiAl alloy phase, as shown in Figure 12. The TiAl alloy powder is a coarse powder bonded to the surface of the TiAl alloy, with a high oxygen content and no Ti element, which can be determined as Al_2O_3 powder. Based on the above analysis, the optimal reaction temperature for the thermal reduction of $AlCl_3$&KCl mixed molten salt magnesium is 750 °C, and the main product generated by the reaction at 750 °C is $TiAl_3$. The oxygen content of the product was analyzed using the JXA82 electric probe. The effect of the reaction time on the oxygen content of the product is shown in Figure 13, with the lowest oxygen content of 3.91 wt% after two hours of reaction.

3.3. Analysis of the $AlCl_3$&KCl-Molten-Salt-Assisted Magnesium Thermal Reduction Process

The above experiments indicate that the $AlCl_3$&KCl molten salt system serves as a medium for the magnesium thermal reduction of TiO_2, to prepare $TiAl_3$ alloy powder. The schematic diagram of the entire reaction process is shown in Figure 14. At the beginning of the reaction, Mg is ionized in the molten salt to form Mg^{2+}, which reacts with Cl^- in the molten salt to form $MgCl_2$, while Al^{3+} is reduced to metallic aluminium. At this time, the metallic aluminium is not covered by a surface oxide film, and the generated liquid metal Al reacts with TiO_2 to generate TiAl-based alloys. The melting temperature of Al is relatively low. At 750 °C, a liquid–solid reaction occurs between the Al liquid and the reduced Ti solid, generating $TiAl_3$ on the surface of the titanium. Afterwards, the internal metal Ti reacts with $TiAl_3$ to generate $TiAl_2$. With the prolongation of the holding time, various TiAl-based alloys react, to generate TiAl alloy powder. The entire TiAl alloy formation process is shown in Equations (5)–(9). This experiment can prepare TiAl₃ alloy powder stably at 750 °C for 2 h, demonstrating a new method for preparing TiAl alloy at low temperatures. Moreover, the TiAl₃ alloy powder has a low density, high modulus, and strong oxidation resistance, making it an excellent high-temperature structural and layer material.

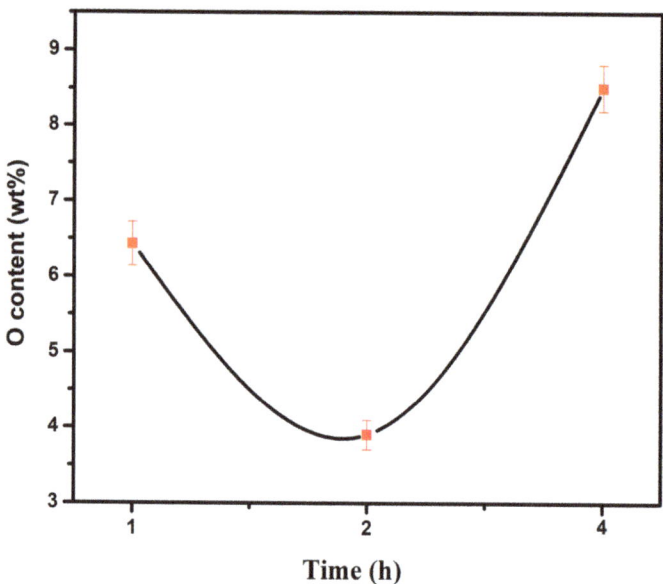

Figure 13. The effect of different reaction times on the oxygen content of products.

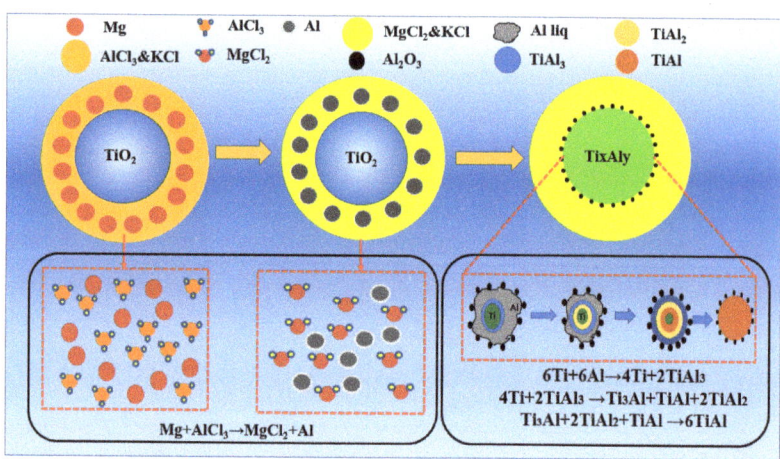

Figure 14. Schematic diagram of the AlCl₃&KCl-molten-salt-assisted magnesium thermal reduction process.

3.4. Product Analysis and Comparison

The elements of the TiAl₃ alloy powder product prepared in this experiment are shown in Table 2, and the particle size is shown in Figure 15. The Ti content is approximately 28.41 wt%, and the Al content is approximately 67.68 wt%. The TiAl₃ alloy prepared in this experiment has a larger particle size. The final products D10, D50, and D90 were

135 μm, 453 μm, and 912 μm, respectively. From the discovery of metallic titanium to the current preparation of titanium alloys, in addition to traditional high-temperature melting and mechanical alloying, many low-cost methods have also been used for preparing TiAl-based alloys. The TiAl-based alloy shows an excellent performance. Developing low-cost and straightforward equipment for new processes, to meet increasing performance requirements, is a hot research topic. These new processes all have the advantages of continuity and low cost. Among emerging technologies, high-temperature self-propagation is the most representative method, but the process controllability of this method is too poor, and there are certain risks. The multi-level deep reduction method is also a relatively novel method that has recently emerged. Still, it also has the problem of a high reaction temperature, which leads to a high energy consumption. Table 3 compares new methods for preparing TiAl-based alloys in recent years, and this experiment. These preparation processes for preparing metallic titanium alloys are currently at the laboratory stage, and there is still a long way to go from the laboratory to industrial production.

Table 2. Composition of the TiAl$_3$ alloy (wt%).

Ti	Al	O
28.41	67.68	3.91

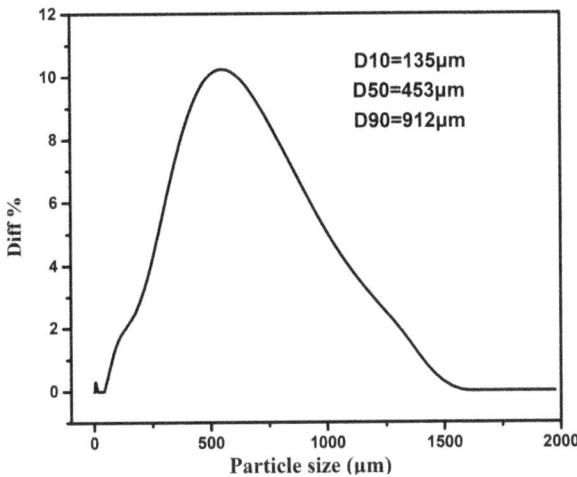

Figure 15. Particle size distribution of the TiAl$_3$ alloy powder.

Table 3. Comparison of new methods for titanium alloy preparation.

Method	Features	Advantages	Disadvantages
Perform reduction process (PRP) [39]	Ca as a reducing agent	Complete reaction with a high recovery rate	Unable to apply on a large scale
SHS process [40]	Spontaneous reaction through the heating agent	Efficiency, high energy consumption, and low consumption	Uncontrollable reaction process
Two-stage thermite process [41]	Reduction using Na$_2$TiF$_6$ as a raw material	Short process and low energy consumption	Multiple reactions
AlCl$_3$&KCl-molten-salt assisted magnesium thermal reduction	Low temperature, AlCl$_3$&KCl as the reaction medium	Low energy consumption, controllable reaction	AlCl$_3$ volatilization

4. Conclusions

This article proposes a new process for preparing $TiAl_3$ alloy powder, mainly discussing how $AlCl_3$&KCl molten salt can prepare $TiAl_3$ alloy powder via the magnesium thermal reduction of TiO_2. At the same time, the influence of different experimental factors on the reduction effect is studied, mainly the influence of the reduction temperature and time on the product phase and oxygen content. Based on the experimental research, the following main conclusions have been drawn:

(1) Through the study of the basic physical properties of the $AlCl_3$ and $AlCl_3$&KCl, it was found that the density of the $AlCl_3$ decreased from 1.3 g·cm^{-3} to 1 g·cm^{-3} in the temperature range of 460–560 °C, and the viscosity also decreased from 0.34 Pa·s to 0.27 Pa·s with the temperature. For $AlCl_3$&KCl eutectic salts, when the KCl content is 20 wt% and 33.33 wt%, the density of the $AlCl_3$&KCl decreases from 1.65 g·cm^{-3} to below 1.5 g·cm^{-3} within the range of 350–850 °C. However, when the KCl content is 50 wt% and 66.66 wt%, the temperature has little effect on the density change of the $AlCl_3$&KCl, with a density of approximately 2 g·cm^{-3}. To reduce the volatilization of the $AlCl_3$, the experiment selected a mass ratio of 0.65 for the $AlCl_3/(AlCl_3+KCl)$ as the selected eutectic salt ratio.

(2) Using TiO_2 as the raw material, magnesium thermal reduction experiments were conducted in a $AlCl_3$&KCl molten salt medium, to clarify the reaction action and sequence of $AlCl_3$ during the reaction process. During the magnesium thermal reduction process, Mg preferentially reacts with $AlCl_3$ at 400–1000 °C to generate metal Al, which reacts with TiO_2, and generates $TiAl_3$ alloy powder at 750–950 °C.

(3) When the experimental temperature is within the range of 750–950 °C, the products of the $AlCl_3$&KCl molten-salt-assisted magnesium thermal reduction gradually form a TiAl alloy from the $TiAl_3$ alloy at 750 °C to 950 °C. The oxygen content also increases from 4.23 wt% at 750 °C, to 11.23 wt% at 950 °C, and the final powder oxygen content also increases with the extension of the reduction time. After 4 h of reaction at 750 °C, the oxygen content reaches approximately 8.5 wt%, and the reaction temperature is at 950 °C; after 4 h of reaction time, it will cause an increase in, and adhesion of, the Al_2O_3 in the product. The optimal reaction time for this process is 2 h, and the reaction temperature is 750 °C. $TiAl_3$ alloy powders with Ti, Al, and O contents of 28.41 wt%, 67.68 wt%, and 3.91 wt% can be obtained, respectively. The powder particle size is concentrated at around 450 μm.

Author Contributions: Software, Y.C.; writing—original draft, J.K.; writing—review and editing, D.Z.; project administration, G.Q. and X.L. All authors have read and agreed to the published version of the manuscript.

Funding: This research was funded by National Natural Science Foundation of China, grant number [52074052].

Data Availability Statement: Not applicable.

Acknowledgments: The authors are incredibly grateful for a grant from the National Natural Science Foundation of China (Grant No.52074052).

Conflicts of Interest: The authors declare no conflict of interest.

References

1. Tetsui, T. Impact Resistance of Commercially Applied TiAl Alloys and Simple-Composition TiAl Alloys at Various Temperatures. *Metals* **2022**, *12*, 2003.
2. Wang, J.H.; Lu, Y.; Shao, X.H. First-Principles Calculation for the Influence of C and O on the Mechanical Properties of gamma-TiAl Alloy at High Temperature. *Metals* **2019**, *9*, 262.
3. Yang, Y.; Liang, Y.F.; Li, C.; Lin, J. Microstructure and Mechanical Properties of TiAl Matrix Composites Reinforced by Carbides. *Metals* **2022**, *12*, 790.
4. Mogale, N.F.; Matizamhuka, W.R. Spark Plasma Sintering of Titanium Aluminides: A Progress Review on Processing, Structure-Property Relations, Alloy Development and Challenges. *Metals* **2020**, *10*, 1080.

5. Wang, Z.H.; Sun, H.X.; Du, Y.L.; Yuan, J. Effects of Powder Preparation and Sintering Temperature on Properties of Spark Plasma Sintered Ti-48Al-2Cr-8Nb Alloy. *Metals* **2019**, *9*, 861.
6. Dong, Z.C.; Feng, A.H.; Wang, H.; Qu, S.; Wang, H. Thermodynamic Study on Initial Oxidation Behavior of TiAl-Nb Alloys at High Temperature. *Metals* **2023**, *13*, 485.
7. Zhu, X.P.; Zhu, C.L.; Lin, B.S.; Wang, Z. Research on Optimization Design of Cast Process for TiAl Case Casting. *Metals* **2022**, *12*, 1954.
8. WilliamsI, J.C.; Boyer, R.R. Opportunities and Issues in the Application of Titanium Alloys for Aerospace Components. *Metals* **2020**, *10*, 705.
9. Zhang, S.L.; Cui, N.; Sun, W.; Li, Q. Microstructural Characterization and Crack Propagation Behavior of a Novel beta-Solidifying TiAl Alloy. *Metals* **2021**, *11*, 1231.
10. Yu, W.; Zhou, J.X.; Yin, Y.; Feng, X.; Nan, H.; Lin, J.; Ding, X.; Duan, W. Effects of Hot Isostatic Pressing and Heat Treatment on the Microstructure and Mechanical Properties of Cast TiAl Alloy. *Metals* **2021**, *11*, 1156.
11. Bewlay, B.P.; Nag, S.; Suzuki, A.; Weimer, M.J. TiAl alloys in commercial aircraft engines. *Mater. High Temp.* **2016**, *33*, 549–559.
12. Galati, M.; Gatto, M.L.; Bloise, N.; Fassina, L.; Saboori, A.; Visai, L.; Mengucci, P.; Iuliano, L. Electron Beam Powder Bed Fusion of Ti-48Al-2Cr-2Nb Open Porous Scaffold for Biomedical Applications: Process Parameters, Adhesion, and Proliferation of NIH-3T3 Cells. *3D Print. Addit. Manuf.* **2022**. [CrossRef]
13. Xu, R.; Li, M.; Zhao, Y. A review of microstructure control and mechanical performance optimization of γ-TiAl alloys. *J. Alloys Compd.* **2023**, *932*, 167611.
14. Subramanyam, R.B. Some recent innovations in the Kroll process of titanium sponge production. *Bull. Mater. Sci.* **1993**, *16*, 433–451.
15. Wang, W.; Wu, F. Quantifying Heat Transfer Characteristics of the Kroll Reactor in Titanium Sponge Production. *Front. Energy Res.* **2021**, *9*, 759781.
16. Wartman, F.S.; Baker, D.H.; Nettle, J.R.; Homme, V.E. Some Observations on the Kroll Process for Titanium. *J. Electrochem. Soc.* **1954**, *101*, 507–513.
17. Zhang, W.; Zhu, Z.; Cheng, C.Y. A literature review of titanium metallurgical processes. *Hydrometallurgy* **2011**, *108*, 177–188.
18. Kumaran, S.; Chantaiah, B.; Rao, T.S. Effect of niobium and aluminium additions in TiAl prealloyed powders during high-energy ball milling. *Mater. Chem. Phys.* **2008**, *108*, 97–101.
19. Rao, K.P.; Du, Y.J. In situ formation of titanium silicides-reinforced TiAl-based composites. *Mater. Sci. Eng. A* **2000**, *277*, 46–56.
20. Wang, G.X.; Mao, X. Preparation of TiAl/Mo and TiAl/NiAl composites by powder processing. *J. Mater. Sci.* **1997**, *32*, 6325–6329.
21. Yan, M.; Yang, F.; Zhang, H.; Zhang, C.; Zhang, H.; Chen, C.; Guo, Z. Multiple intermetallic compounds reinforced Ti–48Al alloy with simple composition and high strength. *Mater. Sci. Eng. A* **2022**, *858*, 144152.
22. Pilone, D.; Pulci, G.; Paglia, L.; Mondal, A.; Marra, F.; Felli, F.; Brotzu, A. Mechanical Behaviour of an Al_2O_3 Dispersion Strengthened γTiAl Alloy Produced by Centrifugal Casting. *Metals* **2020**, *10*, 1457.
23. Shen, Y.; Jia, Q.; Zhang, X.; Liu, R.; Wang, Y.; Cui, Y.; Yang, R. Tensile Behavior of SiC Fiber-Reinforced γ-TiAl Composites Prepared by Suction Casting. *Acta Metall. Sin.* **2021**, *34*, 932–942.
24. Tetsui, T. Selection of Additive Elements Focusing on Impact Resistance in Practical TiAl Cast Alloys. *Metals* **2022**, *12*, 544.
25. Yu, Y.; Kou, H.; Wang, Y.; Wang, Y.; Jia, M.; Li, H.; Li, Y.; Wang, J.; Li, J. Controlling lamellar orientation of Ti-47.5Al-5Nb-2.5V-1Cr alloy by conventional casting. *Scr. Mater.* **2023**, *223*, 115080.
26. Lapin, J.; Kamyshnykova, K.; Klimova, A. Comparative Study of Microstructure and Mechanical Properties of Two TiAl-Based Alloys Reinforced with Carbide Particles. *Molecules* **2020**, *25*, 3423.
27. Gabrisch, H.; Stark, A.; Schimansky, F.-P.; Wang, L.; Schell, N.; Lorenz, U.; Pyczak, F. Investigation of carbides in Ti–45Al–5Nb–xC alloys ($0 \leq x \leq 1$) by transmission electron microscopy and high energy-XRD. *Intermetallics* **2013**, *33*, 44–53.
28. Emiralioğlu, A.; Ünal, R. Additive manufacturing of gamma titanium aluminide alloys: A review. *J. Mater. Sci.* **2022**, *57*, 4441–4466.
29. Liu, J.; Wang, M.; Zhang, P.; Chen, Y.; Wang, S.; Wu, T.; Xie, M.; Wang, L.; Wang, K. Texture refinement and mechanical improvement in beam oscillation superimposed laser welding of TiAl-based alloy. *Mater. Charact.* **2022**, *188*, 111892.
30. Liu, Z.-Q.; Zhu, X.-O.; Yin, G.-L.; Zhou, Q. Direct bonding of bimetallic structure from Ti_6Al_4V to $Ti_{48}Al_2Cr_2Nb$ alloy by laser additive manufacturing. *Mater. Sci. Technol.* **2022**, *38*, 39–44.
31. Soliman, H.A.; Elbestawi, M. Titanium aluminides processing by additive manufacturing—A review. *Int. J. Adv. Manuf. Technol.* **2022**, *119*, 5583–5614.
32. Zhou, W.; Shen, C.; Hua, X.; Wang, L.; Zhang, Y.; Li, F.; Xin, J.; Ding, Y. The effect of vanadium on the microstructure and mechanical properties of TiAl alloy fabricated by twin-wire directed energy deposition-arc. *Addit. Manuf.* **2023**, *62*, 103382.
33. Zhao, K.; Feng, N.; Wang, Y. Fabrication of Ti-Al intermetallics by a two-stage aluminothermic reduction process using Na_2TiF_6. *Intermetallics* **2017**, *85*, 156–162.
34. Zhao, K.; Wang, Y.W.; Peng, J.P. Electrochemical preparation of titanium and titanium-copper alloys with $K_2Ti_6O_{13}$ in KF-KCl melts. *Rare Met.* **2017**, *36*, 527–532.
35. Song, Y.; Dou, Z.; Zhang, T.-A. Mechanisms of metal-slag separation behavior in thermite reduction for preparation of TiAl alloy. *J. Mater. Eng. Perform.* **2021**, *30*, 9315–9325.
36. Song, Y.; Dou, Z.; Liu, Y.; Zhang, T.-A. Study on the Preparation Process of TiAl Alloy by Self-Propagating Metallurgy. *J. Mater. Eng. Perform.* **2023**. [CrossRef]

37. Janz, G.J.; Tomkins, R.P.T.; Allen, C.B.; Downey, J.R.; Garner, G.L.; Krebs, U.; Singer, S.K. Molten salts: Volume 4, part 2, chlorides and mixtures—Electrical conductance, density, viscosity, and surface tension data. *J. Phys. Chem. Ref. Data* **1975**, *4*, 871–1178.
38. Školáková, A.; Leitner, J.; Salvetr, P.; Novák, P.; Deduytsche, D.; Kopeček, J.; Detavernier, C.; Vojtěch, D. Kinetic and thermodynamic description of intermediary phase formation in Ti-Al system during reactive sintering. *Mater. Chem. Phys.* **2019**, *230*, 122–130. [CrossRef]
39. Song, Y.L.; Dou, Z.H.; Zhang, T.A.; Liu, Y. Research Progress on the Extractive Metallurgy of Titanium and Its Alloys. *Miner. Process. Extr. Metall. Rev.* **2020**, *42*, 535–551.
40. Choi, K.; Choi, H.; Sohn, I. Understanding the magnesiothermic reduction mechanism of TiO_2 to produce Ti. *Metall. Mater. Trans. B* **2017**, *48*, 922–932.
41. Zhao, K.; Wang, Y.W.; Feng, N.X. Cleaner production of Ti powder by a two-stage aluminothermic reduction process. *JOM* **2017**, *69*, 1795–1800.

Disclaimer/Publisher's Note: The statements, opinions and data contained in all publications are solely those of the individual author(s) and contributor(s) and not of MDPI and/or the editor(s). MDPI and/or the editor(s) disclaim responsibility for any injury to people or property resulting from any ideas, methods, instructions or products referred to in the content.

Article

First-Principle Investigation into Mechanical Properties of Al$_6$Mg$_1$Zr$_1$ under Uniaxial Tension Strain on the Basis of Density Functional Theory

Lihua Zhang [1,†], Jijun Li [2,*,†], Jing Zhang [2], Yanjie Liu [2,*] and Lin Lin [3,*]

1 School of Science, Shanghai Maritime University, Shanghai 201306, China; zhanglh@shmtu.edu.cn
2 School of Mechanical and Energy Engineering, Shanghai Technical Institute of Electronics and Information, Shanghai 201411, China
3 College of Science, Inner Mongolia University of Technology, Hohhot 010051, China
* Correspondence: lijijun@stiei.edu.cn (J.L.); liuyanjie@stiei.edu.cn (Y.L.); linlin@imut.edu.cn (L.L.)
† First authors.

Abstract: The influences of uniaxial tension strain in the x direction (ε_x) on the mechanical stability, stress–strain relations, elastic properties, hardness, ductility, and elastic anisotropy of Al$_6$Mg$_1$Zr$_1$ compound were studied by performing first-principle calculations on the basis of density functional theory. It was found that Al$_6$Mg$_1$Zr$_1$ compound is mechanically stable in the range of strain ε_x from 0 to 6%. As the strain ε_x increased from 0 to 6%, the stress in the x direction (σ_x) first grew linearly and then followed a nonlinear trend, while the stresses in the y and z directions (σ_y and σ_z) showed a linearly, increasing trend all the way. The bulk modulus B, shear modulus G, and Young's modulus E all dropped as the strain ε_x increased from 0 to 6%. The Poisson ratio μ of Al$_6$Mg$_1$Zr$_1$ compound was nearly unchanged when the strain ε_x was less than 3%, but then it grew quickly. Vickers hardness H_V of Al$_6$Mg$_1$Zr$_1$ compound dropped gradually as the strain ε_x increased from 0 to 6%. The Al$_6$Mg$_1$Zr$_1$ compound was brittle when the ε_x was less than 4%, but it presented ductility when the strain ε_x was more than 4%. As the strain ε_x increased from 0 to 6%, the compression anisotropy percentage (A_B) grew and its slope became larger when the strain ε_x was more than 4%, while both the shear anisotropy percentage (A_G) and the universal anisotropy index (A_U) first dropped slowly and then grew quickly. These results demonstrate that imposing appropriate uniaxial tension strain can affect and regulate the mechanical properties of Al$_6$Mg$_1$Zr$_1$ compound.

Keywords: Al$_6$Mg$_1$Zr$_1$; mechanical properties; uniaxial tension strain; first principles

Citation: Zhang, L.; Li, J.; Zhang, J.; Liu, Y.; Lin, L. First-Principle Investigation into Mechanical Properties of Al$_6$Mg$_1$Zr$_1$ under Uniaxial Tension Strain on the Basis of Density Functional Theory. *Metals* **2023**, *13*, 1569. https://doi.org/10.3390/met13091569

Academic Editors: Alain Pasturel and Varvara Romanova

Received: 5 July 2023
Revised: 14 August 2023
Accepted: 29 August 2023
Published: 7 September 2023

Copyright: © 2023 by the authors. Licensee MDPI, Basel, Switzerland. This article is an open access article distributed under the terms and conditions of the Creative Commons Attribution (CC BY) license (https://creativecommons.org/licenses/by/4.0/).

1. Introduction

From an industrial point of view, the aluminum–magnesium (Al-Mg) based alloys and intermetallic compounds are considered as very promising engineering materials, and have been widely used in aviation, aerospace, shipbuilding, rail transportation, and automotive industries owing to their high strength-to-weight ratio, good formability, excellent corrosion resistance, and good weldability [1–4]. However, Al-Mg based materials only have a low to medium strength, which severely restricts their application in industry [5–7]. With the rapid development of industrial technology, a further improvement of the comprehensive performance of Al-Mg based materials is required. An effective approach to improve the performance of the Al-Mg based materials is by adding the appropriate elements [8–12]. The rare-earth metal scandium (Sc) has proved to be the most effective element to improve the performance of the Al-Mg based materials [13–16]. However, the application of Sc is greatly limited in industry due to its high cost. Therefore, it is of great significance to search for another lower-cost additional element to replace Sc. The transition metal zirconium (Zr) is of much lower cost (its price is only about 1/100 of that of Sc), and can serve a similar strengthening function to Sc in Al-Mg based materials [17–21].

In recent years, it was reported that the functional properties of materials were regulated by imposing strain [22–25]. Fu, et al. quantitatively analyzed the effects of heterogeneous plastic strain on hydrogen-induced cracking of twinning-induced plasticity (TWIP) steel by electron backscattered diffraction (EBSD) technology. According to a quantitative calculation of EBSD crystallographic data, the larger the geometrically necessary dislocation density (ρ_{GND}), the greater is the heterogeneous strain. It was evident from the overall statistical analyses that hydrogen-induced crack initiation depends on interactions between heterogeneous strain and hydrogen atoms generated by local grain orientation deviation (deviation from ideal) and gradient (misorientation). The larger the ρ_{GND}, the easier it is to cause local enrichment of hydrogen atoms, which in turn leads to hydrogen-induced cracking in TWIP steel [26]. Liang et al. investigated the strain-induced strengthening in superconducting β-Mo_2C through high pressure and high temperature. It was found that strain-induced high-density dislocations and low-angle grain boundaries were introduced and enabled the synthesized β-Mo_2C ceramics to exhibit surprising mechanical properties [27]. Du, et al. studied the Poisson ratio of in-plane pristine armchair and zigzag graphene under uniaxial tensile loading by molecular dynamics simulations, which indicated that the Poisson ratio strongly depends on the tensile strain. At the critical strain, the Poisson ratio will transform from positive to negative, and the critical strain of the zigzag is far less than that of armchair [28]. Rasidul Islam, et al. investigated the strain-induced mechanical properties of inorganic halide perovskite $CsGeBr_3$ through first-principles based on density functional theory. It was found that the bulk modulus, shear modulus, and Young's modulus all increased with increasing compressive strain but decreased with increasing tensile strain. The brittleness of $CsGeBr_3$ increased with compressive strain, whereas CsGeBr3 offered significant ductility with more than 2% tensile strain [29]. Tan et al. investigated the mechanical behavior of $AlSi_2Sc_2$ under uniaxial tensile strain by performing first-principle calculations based on density functional theory. It was found that the estimated elastic moduli of $AlSi_2Sc_2$ decreased with increasing uniaxial tensile strain, but the brittleness of $AlSi_2Sc_2$ did not change when strain was applied [30]. However, there has been little investigation into the effect of strain on the mechanical properties of Al-Mg-Zr compounds based on the density functional theory.

In the present study, the first principle calculations based on density functional theory were used to investigate the elastic properties, hardness, ductility, and elastic anisotropy of $Al_6Mg_1Zr_1$ compound at different uniaxial tensile strains. The current research will contribute to a better understanding of the modulation of strain on the mechanical properties of Al-Mg based alloys and compounds.

2. Computational Methods

In this study, the structural model of $Al_6Mg_1Zr_1$ supercell viewed along the c axis is shown in Figure 1. Grey, orange, and green spheres represent Al, Mg, and Zr atoms, respectively. The x, y, and z direction are parallel to the a, b, and c axis, respectively. Our calculations of $Al_6Mg_1Zr_1$ were carried out by the Cambridge Serial Total Energy Package (CASTEP) code, using the plane-wave pseudopotential method based on density functional theory (DFT) [31–33]. For the exchange and correlation terms in the electron–electron interaction, the generalized gradient approximation (GGA) in the scheme of Perdew–Burke–Ernzerhof (PBE) was used [34]. The valence wave functions were expanded in a plane-wave basis set up to an energy cutoff of 600 eV. For the k point sampling, a $3 \times 6 \times 6$ Monkhorst–Pack mesh in the Brillouin zone was used. The other parameters used default settings of ultra-fine accuracy.

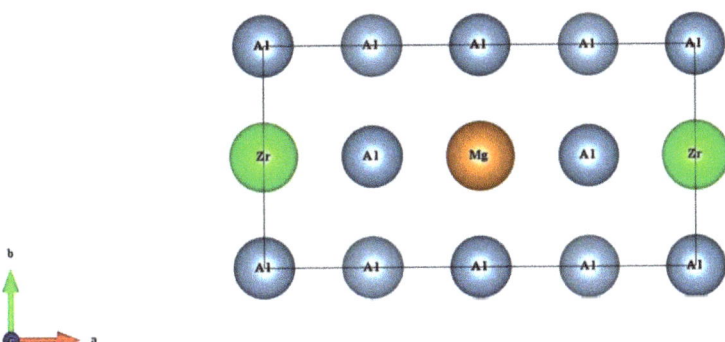

Figure 1. Structural model of $Al_6Mg_1Zr_1$ supercell.

3. Results and Discussion

3.1. Mechanical Stability

Elastic stiffness constants C_{ij} are very important physical quantities used to express the elasticity of a solid material in engineering applications. In this work, the elastic stiffness constants were calculated using the stress–strain approach based on Hook's law by imposing uniaxial tension strain in the x direction (ε_x). The elastic stiffness constants C_{ij} of $Al_6Mg_1Zr_1$ compound at different ε_x are presented in Table 1. It can be found that there are nine independent effective constants (C_{11}, C_{12}, C_{13}, C_{22}, C_{23}, C_{33}, C_{44}, C_{55}, and C_{66}) in the elastic stiffness matrix for $Al_6Mg_1Zr_1$. Therefore, $Al_6Mg_1Zr_1$ is determined to be of orthorhombic structure [35].

Table 1. Calculated values of the elastic stiffness constants C_{ij} (in GPa) of $Al_6Mg_1Zr_1$ at different uniaxial tension strains in x direction (ε_x).

ε_x (%)	C_{11}	C_{12}	C_{13}	C_{22}	C_{23}	C_{33}	C_{44}	C_{55}	C_{66}
0	157.17	41.58	41.56	170.16	41.57	170.08	40.52	53.70	53.71
1%	146.98	36.34	36.37	163.72	42.30	163.79	42.19	49.69	49.90
2%	136.71	35.30	35.32	156.80	45.54	156.82	44.84	47.06	47.06
3%	124.44	35.59	35.57	146.39	48.13	146.36	48.54	43.25	43.25
4%	111.37	37.46	37.46	133.26	50.68	133.26	51.02	37.65	37.65
5%	94.62	39.78	39.78	118.19	55.44	118.19	50.42	29.91	29.91
6%	65.16	44.63	44.63	99.25	64.20	99.26	46.10	19.98	19.98
7%	17.81	52.18	52.20	75.93	75.41	75.95	40.10	8.17	8.17

Based on Born–Huang's dynamical theory of crystal lattices, the mechanical stability standards for orthorhombic crystals must meet the following requirements [36]:

$$\begin{cases} C_{ii} > 0 \\ C_{11} + C_{22} - 2C_{12} > 0 \\ C_{11} + C_{33} - 2C_{13} > 0 \\ C_{22} + C_{33} - 2C_{23} > 0 \\ C_{11} + C_{22} + C_{33} + 2(C_{12} + C_{13} + C_{23}) > 0 \end{cases} \quad (1)$$

It was noted that the elastic constants C_{ij} of the $Al_6Mg_1Zr_1$ fulfilled well the mechanical stability standards in the range of strain ε_x from 0 to 6%, while they could not meet the standards when the ε_x was more than 6%. Therefore, the orthorhombic of $Al_6Mg_1Zr_1$ is mechanically stable in the range of strain ε_x from 0 to 6%. In this study, we are only concerned with the mechanical properties of $Al_6Mg_1Zr_1$ in the range of strain ε_x from 0 to 6%.

3.2. Stress–Strain Relations

The stresses in the principal axis direction (σ_x, σ_y, and σ_z) of $Al_6Mg_1Zr_1$ at different uniaxial tension strains in the x direction (ε_x) were calculated based on Hook's law, and the calculated values are presented in Table 2 and Figure 2.

Table 2. Calculated values of stresses in principal axis directions (σ_x, σ_y, and σ_z) of $Al_6Mg_1Zr_1$ at different uniaxial tension strains in x direction (ε_x).

ε_x (%)	σ_x (GPa)	σ_y (GPa)	σ_z (GPa)
0	0	0	0
1%	1.57	0.42	0.42
2%	3.04	0.78	0.78
3%	4.41	1.13	1.13
4%	5.65	1.49	1.49
5%	6.77	1.87	1.86
6%	7.71	2.26	2.26

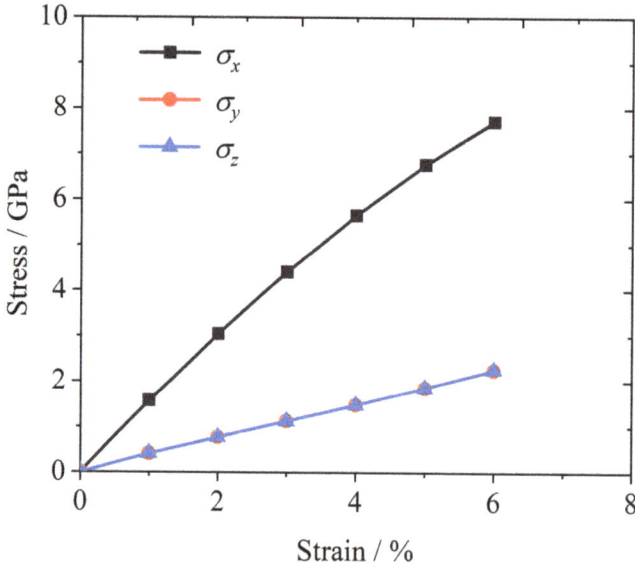

Figure 2. Stress–strain relations of $Al_6Mg_1Zr_1$ at different uniaxial tension strains in x direction (ε_x).

It can be seen that, as the strain ε_x increased from 0 to 6%, the stress in the x direction (i.e., σ_x) gradually grew to 7.7129 GPa. The σ_x-ε_x curve of $Al_6Mg_1Zr_1$ had a linear formation in the range of strain ε_x from 0 to 3%, and then followed a nonlinear trend in the range of strain ε_x from 3% to 6%. It is indicated that the deformation of $Al_6Mg_1Zr_1$ was elastic in the range of ε_x from 0 to 3%, after which plastic deformation occurred. The stresses in the y and z directions (σ_y and σ_z) were almost equal and had good linear relation with the uniaxial tension strain in the range of ε_x from 0 to 6%. On the whole, the stress σ_x was much higher than σ_y and σ_z due to the uniaxial tension loading being in the x direction.

3.3. Elastic Properties of Polycrystalline Materials

In many cases, polycrystalline materials have advantages in practical applications compared to single crystal materials [37]. Therefore, it is more meaningful to examine the elastic properties of polycrystalline materials. The elastic properties of polycrystalline materials can be characterized by the bulk modulus B, shear modulus G, Young's modulus E, and Poisson ratio μ.

There are two approximation methods to obtain polycrystalline elastic moduli, namely, the Voigt method and the Reuss method. The Voigt method provides the upper bound to the polycrystalline elastic moduli, and the Reuss method provides the lower bound to the polycrystalline elastic moduli. For different crystalline systems, the bulk modulus B and shear modulus G according to Voigt and Reuss approximations are given by the following equations [38]:

$$B_V = \frac{C_{11} + C_{22} + C_{33} + 2(C_{12} + C_{13} + C_{23})}{9} \quad (2)$$

$$G_V = \frac{C_{11} + C_{22} + C_{33} - C_{12} - C_{13} - C_{23}}{15} + \frac{C_{44} + C_{55} + C_{66}}{5} \quad (3)$$

$$B_R = \frac{1}{S_{11} + S_{22} + S_{33} + 2(S_{12} + S_{13} + S_{23})} \quad (4)$$

$$G_R = \frac{15}{4(S_{11} + S_{22} + S_{33}) + 3(S_{44} + S_{55} + S_{66}) - 4(S_{12} + S_{13} + S_{23})} \quad (5)$$

where the subscripts V and R denote the Voigt and Reuss averages, C_{ij} are the elastic stiffness constants, and S_{ij} are the elastic compliance coefficients.

The arithmetic average of the Voigt and the Reuss bounds is referred to as the Voigt–Reuss Hill (VRH) average, and it is considered as the best estimate of the theoretical polycrystalline elastic moduli. The VRH averages of B and G are given as follows [38]:

$$B = \frac{B_V + B_R}{2} \quad (6)$$

$$G = \frac{G_V + G_R}{2} \quad (7)$$

The Young's modulus E and Poisson ratio μ of the polycrystalline material can be obtained from the bulk modulus B and shear modulus G, while the corresponding calculation formulas are as follows [38]:

$$E = \frac{9BG}{3B + G} \quad (8)$$

$$\mu = \frac{3B - 2G}{6B + 2G} \quad (9)$$

The Bulk modulus B, shear modulus G, Young's modulus E and Poisson ratio μ of polycrystalline $Al_6Mg_1Zr_1$ at different uniaxial tension strains in the x direction (ε_x) were calculated using the above formulas, and the calculated values are presented in Table 3 and Figure 3.

Table 3. Calculated values of the elastic moduli (B, G, and E) and Poisson ratios μ of polycrystalline $Al_6Mg_1Zr_1$ at different uniaxial tension strains in x direction (ε_x).

ε_x (%)	B (GPa)	G (GPa)	E (GPa)	μ
0	82.93	53.77	132.63	0.23
1%	78.12	51.81	127.28	0.23
2%	75.55	49.82	122.53	0.230
3%	72.43	46.72	115.36	0.24
4%	69.37	41.76	104.34	0.25
5%	65.94	34.36	87.81	0.28
6%	60.63	22.88	60.96	0.33

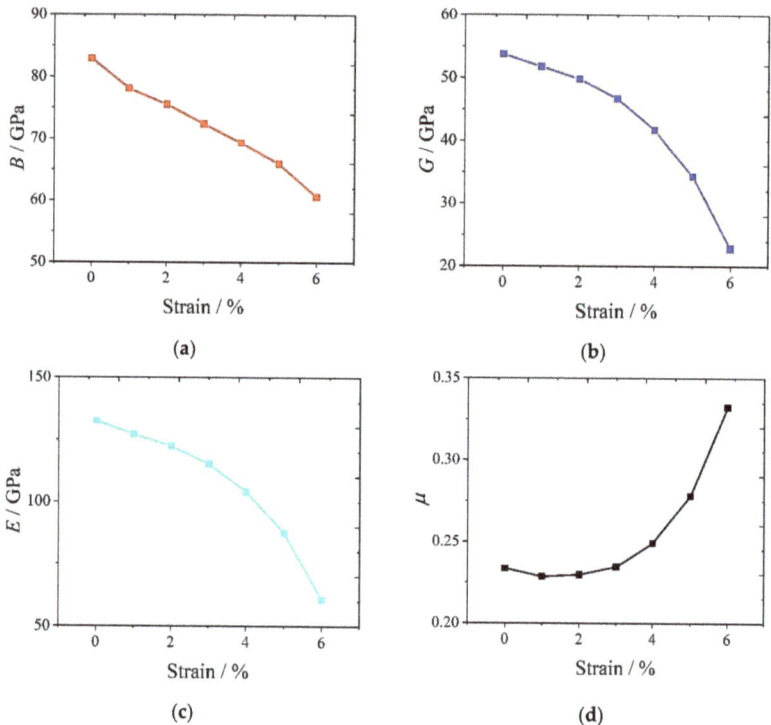

Figure 3. Elastic moduli (B, G, and E) and Poisson ratio μ of Al$_6$Mg$_1$Zr$_1$ at different uniaxial tension strains (ε_x): (**a**) Bulk modulus B vs. strain ε_x; (**b**) Shear modulus G vs. strain ε_x; (**c**) Young's modulus E vs. strain ε_x; (**d**) Poisson ratio μ vs. strain ε_x.

The bulk modulus B is a measure of the resistance of a solid material to a volume change. Figure 3a presents the calculated bulk modulus B of Al$_6$Mg$_1$Zr$_1$ at different strains ε_x. It can be seen that, as the strain ε_x increased from 0 to 6%, the bulk modulus B dropped from 82.93 GPa to 60.63 GPa. The bulk modulus B reduced by 26.9%, which showed its negative relation with the uniaxial tension strain for Al$_6$Mg$_1$Zr$_1$. The Al$_6$Mg$_1$Zr$_1$ alloy has the largest incompressibility at the unstrained state due to the largest B value, while it has the largest compressibility at the strain ε_x of 6% due to the smallest B value.

The shear modulus G is defined as the ratio of shear stress to the shear strain, which characterizes the ability of a solid material to resist deformation under shear stress. The greater G corresponds to the stronger shear resistance of the solid material. Figure 3b presents the calculated values of shear modulus G of Al$_6$Mg$_1$Zr$_1$ at different strains ε_x. It can be seen that as the strain ε_x increased from 0 to 6%, the shear modulus G dropped from 53.77 GPa to 22.88 GPa. The shear modulus G was reduced by 57.5%, which indicated that shear resistance is greatly influenced by the uniaxial tension strain. Al$_6$Mg$_1$Zr$_1$ alloy has the smallest shear resistance at the strain ε_x of 6% due to the smallest G value.

Young's modulus E is defined as the ratio of tensile stress and axial strain and serves as a measure of the stiffness of solid materials. Figure 3c presents the calculated values of Young's modulus E of Al$_6$Mg$_1$Zr$_1$ at different strains ε_x. It can be found that Young's modulus E dropped with increasing strain ε_x. When the Al$_6$Mg$_1$Zr$_1$ was unstrained, the Young's modulus E was 132.63 GPa. When the strain ε_x reached 6%, Young's modulus E dropped to 60.96 GPa. Young's modulus E was reduced by 54.0%, showing its negative relation with uniaxial tension strain. The unstrained Al$_6$Mg$_1$Zr$_1$ has the largest stiffness due to the largest E value, while $Al_6Mg_1Zr_1$ alloy has the smallest stiffness at the strain ε_x of 6% due to the smallest E value.

Figure 3d presents the calculated values of Poisson ratio μ of $Al_6Mg_1Zr_1$ at different strains ε_x. It can be found that when the strain ε_x was less than 3%, the Poisson ratio μ remained nearly unchanged. When ε_x was more than 3%, the Poisson ratio μ grew quickly with increasing strain ε_x, showing its positive relation with uniaxial tension strain. Generally, when the Poisson ratio μ is between -1 and 0.5, the solid is relatively stable under shear deformation. From Figure 3d, it can be seen that the Poisson ratio μ of $Al_6Mg_1Zr_1$ ranged from 0.23 to 0.33, which is between -1 and 0.5, indicating that $Al_6Mg_1Zr_1$ is a stable linear elastic solid. $Al_6Mg_1Zr_1$ has the maximum μ value at the strain ε_x of 6%, indicating that $Al_6Mg_1Zr_1$ has the highest toughness at the strain ε_x of 6%.

By comparing Figures 3a–c and 3d, it can be found that with the strain ε_x increasing from 0 to 6%, the elastic moduli (B, G, and E) monotonically decreased with the increasing strain ε_x, while the Poisson ratio μ was first nearly unchanged and then grew quickly. The variation trends of elastic moduli (B, G, and E) and Poisson ratio μ of $Al_6Mg_1Zr_1$ alloy with the uniaxial tensile strain ε_x are similar to that of $AlSi_2Sc_2$ [30].

3.4. Hardness and Ductility

Hardness is a measure of the resistance to localized deformation induced by either mechanical indentation or abrasion. In general, hardness is linked with the elastic and plastic properties of a material, and the shear modulus G is the more important parameter governing hardness than the bulk modulus B. There have been some semi-empirical models developed to predict the hardness of materials. Chen et al. [39] proposed a model to predict the Vickers hardness H_V of polycrystalline materials and bulk metallic glasses based on the shear modulus G and the Pugh's modulus ratio k ($k = G/B$) as follows [39]:

$$H_V = 1.887 k^{1.171} G^{0.591} \tag{10}$$

The above formula has been successfully used to predict the hardness of many compounds.

The ductility or brittleness of a solid material can be estimated by the ratio of the bulk modulus B to the elastic shear modulus G (i.e., B/G). If the modulus ratio B/G is more than 1.75, the solid is classified as ductile material; otherwise, it is brittle [40,41].

Table 4 and Figure 4 present the calculated values of Vickers hardness H_V and the modulus ratio B/G of $Al_6Mg_1Zr_1$ at different uniaxial tension strains ε_x.

Table 4. Calculated values of the Vickers hardness H_V and the modulus ratio B/G of $Al_6Mg_1Zr_1$ at different uniaxial tension strains in x direction (ε_x).

ε_x (%)	H_V (GPa)	B/G
0	11.97	1.54
1%	12.03	1.51
2%	11.67	1.52
3%	10.95	1.55
4%	9.45	1.66
5%	7.11	1.92
6%	3.83	2.65

From Figure 4a, as the strain ε_x increased from 0 to 6%, the Vickers hardness H_V of $Al_6Mg_1Zr_1$ dropped gradually from 11.97 GPa to 3.83 Gpa. The Vickers hardness H_V of $Al_6Mg_1Zr_1$ dropped by 71.746%, showing its negative relation with the uniaxial tension strain.

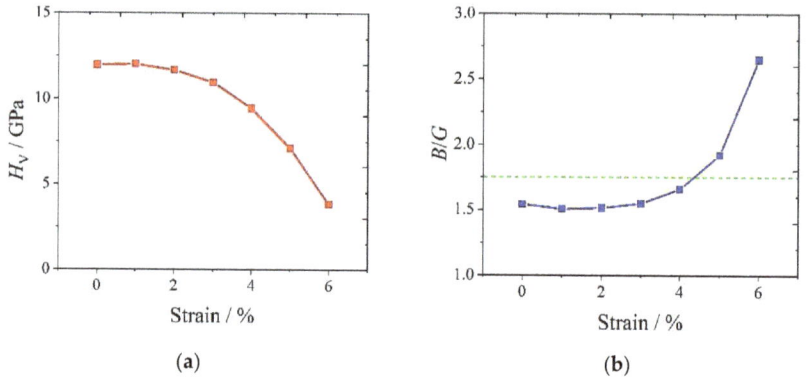

Figure 4. Vickers hardness H_v and the modulus ratio B/G of $Al_6Mg_1Zr_1$ at different uniaxial tension strain (ε_x): (**a**) Vickers hardness H_v vs. strain ε_x; (**b**) the ratio D ($D = B/G$) vs. strain ε_x.

From Figure 4b, when the strain ε_x was less than 3%, the modulus ratio B/G remained nearly unchanged. When the ε_x was more than 3%, the modulus ratio B/G grew quickly from 1.55 to 2.65 with the increasing strain ε_x. When the strain ε_x was less than 4%, the modulus ratio $B/G < 1.75$ and the $Al_6Mg_1Zr_1$ was brittle. When the ε_x was more than 4%, the modulus ratio $B/G > 1.75$ and the $Al_6Mg_1Zr_1$ was taken as ductile. $Al_6Mg_1Zr_1$ would become most ductile at the strain of 6% due to the largest modulus ratio B/G value of 2.65.

3.5. Elastic Anisotropy

Elastic anisotropy of a solid material is very important in diverse applications such as phase transformations and dislocation dynamics, and it can be characterized by the elastic anisotropy indexes. The elastic anisotropy indexes include compression anisotropy percentage A_B, shear anisotropy percentage A_G, and the universal anisotropy index A_U, and can be calculated using the following expressions [42]:

$$\begin{cases} A_B = \dfrac{B_V - B_R}{B_V + B_R} \\ A_G = \dfrac{G_V - G_R}{G_V + G_R} \\ A_U = 5\dfrac{G_V}{G_R} + \dfrac{B_V}{B_R} - 6 \end{cases} \quad (11)$$

If $A_U = A_B = A_G = 0$, the material shows characteristics of elastic isotropy. Otherwise, it has elastic anisotropy, and the larger the deviation of elastic anisotropy index values from 0 (the corresponding elastic isotropy value), the greater its degree in elastic anisotropy.

Table 5 and Figure 5 show the calculated values of elastic anisotropy indexes (A_B, A_G, and A_U) of $Al_6Mg_1Zr_1$ at different uniaxial tension strains ε_x.

Table 5. Calculated values of elastic anisotropy indexes (A_B, A_G, and A_U) of $Al_6Mg_1Zr_1$ at different uniaxial tension strains in x direction (ε_x).

ε_x (%)	A_B	A_G	A_U
0	0.06%	1.26%	0.13
1%	0.21%	0.92%	0.10
2%	0.40%	0.50%	0.06
3%	0.60%	0.32%	0.04
4%	0.77%	0.77%	0.09
5%	1.27%	2.20%	0.25
6%	4.58%	7.37%	0.89

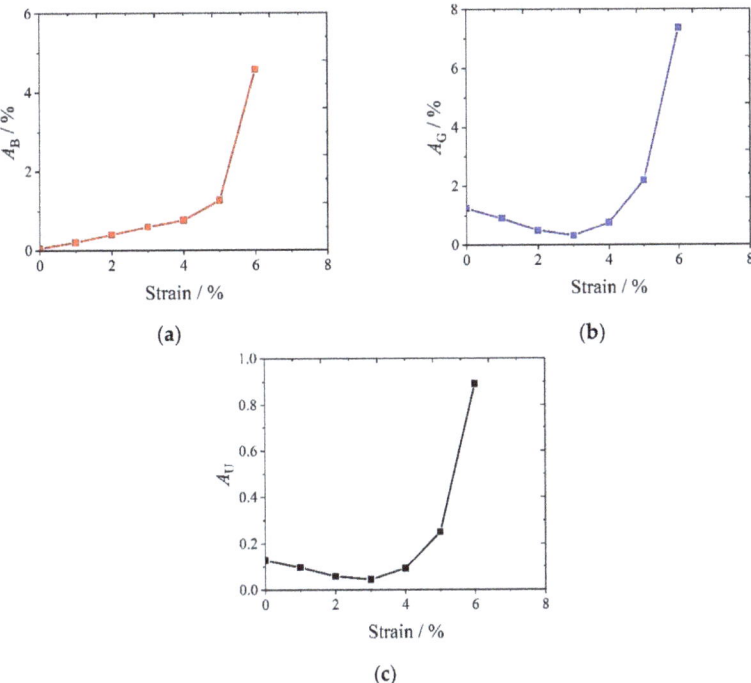

Figure 5. Elastic anisotropy indexes of $Al_6Mg_1Zr_1$ (A_B, A_G, and A_U) at different uniaxial tension strains (ε_x): (**a**) A_B vs. strain ε_x; (**b**) A_G vs. strain ε_x; (**c**) A_U vs. strain ε_x.

From Figure 5a, the compression anisotropy percentage A_B of $Al_6Mg_1Zr_1$ grew slowly as the strain ε_x increased from 0 to 4%, and then it grew quickly as the strain ε_x increased from 4% to 6%. A_B reached its maximum of 4.58% at the strain ε_x of 6%. From Figure 5b, when the strain ε_x was less than 3%, the shear anisotropy percentage A_G dropped slowly with the strain ε_x. A_G reached its minimum of 0.32% at the strain ε_x of 3%. When the strain ε_x was more than 3%, A_G grew quickly with the strain ε_x, and reached its maximum of 7.37% at the strain ε_x of 6%. The universal anisotropy index A_U can reflect the anisotropy more accurately because both bulk modulus B and shear modulus G are deliberated in its expression. As shown in Figure 5c, the change trend of A_U with the strain ε_x is similar to that of A_G. The variation trends of elastic anisotropy indexes of $Al_6Mg_1Zr_1$ with uniaxial tensile strain are similar to that of hexagonal C40 $MoSi_2$ [43]. Moreover, the values of the elastic anisotropy indexes (A_B, A_G, and A_U) are small and the degree in elastic anisotropy of $Al_6Mg_1Zr_1$ is relatively weak in the range of strain ε_x from 0 to 6%.

4. Conclusions

To sum up, in this work the mechanical stability, stress–strain relations, elastic properties, hardness, ductility, and elastic anisotropy of $Al_6Mg_1Zr_1$ at different uniaxial tension strains in the x direction (ε_x) were examined by first principle calculations based on density functional theory. The influences of strain ε_x on the mechanical properties of the $Al_6Mg_1Zr_1$ were studied. It was found that $Al_6Mg_1Zr_1$ was mechanically stable in the range of strain ε_x from 0 to 6%, while it was unstable when the strain ε_x was more than 6%. As the strain ε_x increased from 0 to 6%, the stress in the x direction (σ_x) first increased linearly and then followed a nonlinear trend, but the stresses in the y and z directions (σ_y and σ_z), which were almost equal, showed a linear, increasing trend all the way. Due to the uniaxial tension loading in the x direction, the stress σ_x was much higher than σ_y and σ_z. The bulk modulus B, shear modulus G and Young's modulus E of $Al_6Mg_1Zr_1$ all dropped with increasing strain ε_x from 0 to 6%, showing their negative relations with uniaxial tension strain. Therefore, the incompressibility, shear resistance, and stiffness of the $Al_6Mg_1Zr_1$ all dropped with increasing uniaxial tension strain. When the strain ε_x was less than 3%, The Poisson ratio μ of $Al_6Mg_1Zr_1$ was nearly unchanged. However, it grew quickly when the ε_x was more than 3%, showing its positive relation with uniaxial tension strain. The Poisson ratio μ ranged from 0.23 to 0.33, which is between -1 and 0.5, indicating that $Al_6Mg_1Zr_1$ is stable linear elastic solid in the range of ε_x from 0 to 6%. $Al_6Mg_1Zr_1$ has the highest toughness at the strain ε_x of 6% due to the maximum μ value. The Vickers hardness H_V of $Al_6Mg_1Zr_1$ dropped gradually with the increasing strain ε_x from 0 to 6%, showing its negative relation with uniaxial tension strain. When the strain ε_x was less than 3%, the modulus ratio B/G of $Al_6Mg_1Zr_1$ remained nearly unchanged but it grew quickly when the ε_x was more than 3%, showing its positive relation with uniaxial tension strain. $Al_6Mg_1Zr_1$ was brittle when the ε_x was less than 4%, while it exhibited ductility when the strain ε_x was more than 4%. The best ductility was achieved for $Al_6Mg_1Zr_1$ alloy at the strain ε_x of 6% due to the maximum B/G value. The compression anisotropy percentage A_B of $Al_6Mg_1Zr_1$ grew slowly as the strain ε_x increased from 0 to 4%, while it grew quickly as the strain ε_x increased from 4% to 6%. Both the shear anisotropy percentage (A_G) and universal anisotropy index (A_U) dropped slowly with increasing strain ε_x from 0 to 3%, and then grew quickly with increasing strain ε_x from 3 to 6%. In addition, the values of the elastic anisotropy indexes (A_B, A_G, and A_U) are small and the degree in elastic anisotropy of $Al_6Mg_1Zr_1$ is relatively weak in the range of ε_x from 0 to 6%. These results show that applying uniaxial tension strain is an effective and promising strategy to improve the mechanical properties of $Al_6Mg_1Zr_1$.

Author Contributions: Conceptualization, J.L. and L.L.; methodology, L.Z. and Y.L.; software, L.Z. and J.Z.; validation, L.Z. and Y.L.; formal analysis, L.Z. and J.L.; investigation, L.Z. and J.L.; resources, J.L. and L.L.; data curation, L.Z. and J.Z.; writing—original draft preparation, L.Z. and J.L.; writing—review and editing, J.Z. and L.L.; visualization, L.Z. and J.Z.; supervision, J.L. and Y.L.; project administration, J.L. and Y.L.; funding acquisition, J.L. and L.L. All authors have read and agreed to the published version of the manuscript.

Funding: This research was funded by the Natural Science Foundation of Inner Mongolia Autonomous Region (Grant Nos. 2022MS01009 and 2018MS01013), the National Natural Science Foundation of China (Grant Nos. 11972221 and 11562016), the College Science Research Project of Inner Mongolia Autonomous Region (Grant No. NJZY22383), and the Key Research Project of Inner Mongolia University of Technology (Grant No. ZZ202016).

Institutional Review Board Statement: Not applicable.

Informed Consent Statement: Not applicable.

Data Availability Statement: Data are contained within the article and can be requested from the corresponding author.

Acknowledgments: The authors acknowledge Mechanics Department of Inner Mongolia University of Technology and the Mathematics Department of Shanghai Maritime University for providing technical support.

Conflicts of Interest: The authors declare no conflict of interest.

References

1. Dolce, D.; Swamy, A.; Hoyt, J.; Choudhury, P. Computing the solid-liquid interfacial free energy and anisotropy of the Al-Mg system using a MEAM potential with atomistic simulations. *Comput. Mater. Sci.* **2023**, *217*, 111901. [CrossRef]
2. Xu, W.; Xin, Y.C.; Zhang, B.; Li, X.Y. Stress corrosion cracking resistant nanostructured Al-Mg alloy with low angle grain boundaries. *Acta Mater.* **2022**, *225*, 117607. [CrossRef]
3. Koju, R.K.; Mishin, Y. Atomistic study of grain-boundary segregation and grainboundary diffusion in Al-Mg alloys. *Acta Mater.* **2020**, *201*, 596–603. [CrossRef]
4. Pariyar, A.; Toth, L.S.; Kailas, S.V.; Peltier, L. Imparting high-temperature grain stability to an Al-Mg alloy. *Scr. Mater.* **2021**, *190*, 141–146. [CrossRef]

5. Zhang, H.T.; Guo, C.; Li, S.S.; Li, B.M.; Nagaumi, H. Influence of cold pre-deformation on the microstructure, mechanical properties and corrosion resistance of Zn-bearing 5xxx aluminum alloy. *J. Mater. Res. Technol.* **2022**, *16*, 1202–1212. [CrossRef]
6. Bakare, F.; Schieren, L.; Rouxel, B.; Jiang, L.; Langan, T.; Kupke, A.; Weiss, M.; Dorin, T. The impact of L12 dispersoids and strain rate on the Portevin-Le-Chatelier effect and mechanical properties of Al–Mg alloys. *Mat. Sci. Eng. A Struct.* **2021**, *811*, 141040. [CrossRef]
7. Yi, G.; Cullen, D.A.; Littrell, K.C.; Golumbfskie, W.; Sundberg, E.; Free, M.L. Characterization of Al–Mg alloy aged at low temperatures. *Metall. Mater. Trans. A* **2017**, *48*, 2040–2050. [CrossRef]
8. Zhang, L.; Liu, C.Y.; Zhang, B.; Huang, H.F.; Xie, H.Y.; Cao, K. Mechanical properties of Al–Mg alloys with equiaxed grain structure produced by friction stir processing. *Mater. Chem. Phys.* **2023**, *294*, 127010. [CrossRef]
9. Guo, C.; Zhang, H.T.; Li, J.H. Influence of Zn and/or Ag additions on microstructure and properties of Al-Mg based alloys. *J. Alloys Compd.* **2022**, *904*, 163998. [CrossRef]
10. Stemper, L.; Tunes, M.A.; Paul, O.; Uggowitzer, P.J.; Pogatscher, S. Age-hardening response of AlMgZn alloys with Cu and Ag additions. *Acta Mater.* **2020**, *195*, 541–554. [CrossRef]
11. Zhu, Z.X.; Jiang, X.X.; Wei, G.; Fang, X.G.; Zhong, Z.H.; Song, K.J.; Han, J.; Jiang, Z.Y. Influence of Zn Content on microstructures, mechanical properties and stress corrosion behavior of AA5083 aluminum alloy. *Acta Metall. Sin.* **2020**, *33*, 1369–1378. [CrossRef]
12. Chen, X.L.; Marioara, C.D.; Andersen, S.J.; Friis, J.; Lervik, A.; Holmestad, R.; Kobayashi, E. Precipitation processes and structural evolutions of various GPB zones and two types of S phases in a cold- rolled Al-Mg-Cu alloy. *Mater. Des.* **2021**, *199*, 109425. [CrossRef]
13. Jiang, L.; Zhang, Z.F.; Bai, Y.L.; Li, S.L.; Mao, W.M. Study on Sc microalloying and strengthening mechanism of Al-Mg alloy. *Crystals* **2022**, *12*, 673. [CrossRef]
14. Zakharov, V.V.; Filatov, Y.A.; Fisenko, I.A. Scandium alloying of aluminum alloys. *Met. Sci. Heat Treat.* **2020**, *62*, 518–523. [CrossRef]
15. Yan, K.; Chen, Z.W.; Lu, W.J.; Zhao, Y.N.; Le, W.; Naseem, S. Nucleation and growth of Al$_3$Sc precipitates during isothermal aging of Al-0.55 wt% Sc alloy. *Mater. Charact.* **2021**, *179*, 111331. [CrossRef]
16. Vo, N.Q.; Dunand, D.C.; Seidman, D.N. Atom probe tomographic study of a friction stir-processed Al-Mg-Sc alloy. *Acta Mater.* **2012**, *60*, 7078–7089. [CrossRef]
17. Deng, Y.; Zhu, X.W.; Lai, Y.; Guo, Y.F.; Fu, L.; Xu, G.F.; Huang, J.W. Effects of Zr/(Sc+Zr) microalloying on dynamic recrystallization, dislocation density and hot workability of Al−Mg alloys during hot compression deformation. *Trans. Nonferr. Metal. Soc.* **2023**, *33*, 668–682. [CrossRef]
18. Croteau, J.R.; Jung, J.G.; Whalen, S.A.; Darsell, J.; Mello, A.; Holstine, D.; Lay, K.; Hansen, M.; Dunand, D.C.; Vo, N.Q. Ultrafine-grained Al-Mg-Zr alloy processed by shear-assisted extrusion with high thermal stability. *Scr. Mater.* **2020**, *186*, 326–330. [CrossRef]
19. Kaiser, M.S.; Rashed, H.M. Precipitation behavior prior to hot roll in the cold-rolled Al-Mg and Al-Mg-Zr alloys. *Kov. Mater.* **2022**, *60*, 247–256.
20. Xu, R.; Li, R.D.; Yuan, T.C.; Zhu, H.B.; Wang, M.B.; Li, J.F.; Zhang, W.; Cao, P. Laser powder bed fusion of Al–Mg–Zr alloy: Microstructure, mechanical properties and dynamic precipitation. *Mat. Sci. Eng. A Struct.* **2022**, *859*, 144181. [CrossRef]
21. Croteau, J.R.; Griffiths, S.; Rossell, M.D.; Leinenbach, C.; Kenel, C.; Jansen, V.; Seidman, D.N. Microstructure and mechanical properties of Al-Mg-Zr alloys processed by selective laser melting. *Acta Mater.* **2018**, *153*, 35–44. [CrossRef]
22. Cho, H.; Lee, B.; Jang, D.; Yoon, J.; Chung, S.; Hong, Y. Recent progress in strain-engineered elastic platforms for stretchable thin-film devices. *Mater. Horiz.* **2022**, *9*, 2053–2075. [CrossRef] [PubMed]
23. Chen, Y.M.; Lei, Y.S.; Li, Y.H.; Yu, Y.G.; Cai, J.Z.; Chiu, M.-H.; Rao, R.; Gu, Y.; Wang, C.; Choi, W.; et al. Strain engineering and epitaxial stabilization of halide perovskites. *Nature* **2020**, *577*, 209–215. [CrossRef] [PubMed]
24. Dang, C.Q.; Chou, J.-P.; Dai, B.; Chou, C.-T.; Yang, Y.; Fan, R.; Lin, W.T.; Meng, F.; Hu, A.; Zhu, J.; et al. Achieving large uniform tensile elasticity in microfabricated diamond. *Science* **2021**, *371*, 76–78. [CrossRef] [PubMed]
25. Steuer, O.; Schwarz, D.; Oehme, M.; Schulze, J.; Mączko, H.; Kudrawiec, R.; Fischer, I.A.; Heller, R.; Hübner, R.; Khan, M.M.; et al. Band-gap and strain engineering in GeSn alloys using post-growth pulsed laser melting. *J. Phys. Condens. Mat.* **2023**, *35*, 055302. [CrossRef] [PubMed]
26. Fu, H.; He, W.H.; Mi, Z.S.; Yan, Y.; Zhang, J.L.; Li, J.X. Effects of heterogeneous plastic strain on hydrogen-induced cracking of twinning-induced plasticity steel: A quantitative approach. *Corros. Sci.* **2022**, *209*, 110782. [CrossRef]
27. Liang, H.; He, R.Q.; Lin, W.T.; Liu, L.; Xiang, X.J.; Zhang, Z.G.; Guan, S.X.; Peng, F.; Fang, L. Strain-induced strengthening in superconducting β-Mo$_2$C through high pressure and high temperature. *J. Eur. Ceram. Soc.* **2023**, *43*, 88–98. [CrossRef]
28. Du, Y.-X.; Zhou, L.-J.; Guo, J.-G. Investigation on micro-mechanism of strain-induced and defect-regulated negative Poisson's ratio of grapheme. *Mater. Chem. Phys.* **2022**, *288*, 126412. [CrossRef]
29. Rasidul Islam, M.; Rayid Hasan Mojumder, M.; Moshwan, R.; Jannatul Islam, A.S.M.; Islam, M.A.; Shizer Rahman, M.; Humaun Kabir, M. Strain-driven optical, electronic, and mechanical properties of inorganic halide perovskite CsGeBr$_3$. *ECS J. Solid State Sci. Technol.* **2022**, *11*, 033001. [CrossRef]
30. Tan, Y.; Ma, L.M.; Wang, Y.S.; Zhou, W.; Wang, X.L.; Guo, F. Mechanical and thermodynamic behaviors of AlSi$_2$Sc$_2$ under uniaxial tensile loading: A first-principles study. *J. Phys. Chem. Solids* **2023**, *174*, 111160. [CrossRef]

31. Clark, S.J.; Segall, M.D.; Pickard, C.J.; Hasnip, P.J.; Probert, M.I.J.; Refson, K.; Payne, M.C. First Principles Methods Using CASTEP. *Z. Krist. Cryst. Mater.* **2005**, *220*, 567–570. [CrossRef]
32. Hohenberg, P.; Kohn, W. Inhomogeneous electron gas. *Phys. Rev.* **1964**, *136*, B864–B871. [CrossRef]
33. Kohn, W.; Sham, L.J. Self-consistent equations including exchange and correlation effects. *Phys. Rev.* **1965**, *140*, A1133–A1138. [CrossRef]
34. Perdew, J.P.; Burke, K.; Ernzerhof, K.M. Generalized gradient approximation made simple. *Phys. Rev. Lett.* **1996**, *77*, 3865–3868. [CrossRef] [PubMed]
35. Yang, J.Z.; Yang, D.F.; Wang, Y.Q.; Quan, X.J.; Li, Y.Y. First principles investigation of elastic and thermodynamic properties of CoSbS thermoelectric material. *J. Solid State Chem.* **2021**, *302*, 122443. [CrossRef]
36. Mouhat, F.; Coudert, F.-X. Necessary and sufficient elastic stability conditions in various crystal systems. *Phys. Rev. B* **2014**, *90*, 224104. [CrossRef]
37. Peng, M.J.; Wang, R.F.; Wu, Y.J.; Yang, A.C.; Duan, Y.H. Elastic anisotropies, thermal conductivities and tensile properties of MAX phases Zr_2AlC and Zr_2AlN: A first-principles calculation. *Vacuum* **2022**, *196*, 110715. [CrossRef]
38. Li, L.H.; Wang, W.L.; Wei, B. First-principle and molecular dynamics calculations for physical properties of Ni-Sn alloy system. *Comput. Mater. Sci.* **2015**, *99*, 274–284. [CrossRef]
39. Chen, X.Q.; Niu, H.Y.; Li, D.Z.; Li, Y.Y. Modeling hardness of polycrystalline materials and bulk metallic glasses. *Intermetallics* **2011**, *19*, 1275–1281. [CrossRef]
40. Pugh, S.F. XCII. Relations between the elastic moduli and the plastic properties of polycrystalline pure metals. *Lond. Edinb. Dublin Philos. Mag. J. Sci.* **1954**, *45*, 823–843. [CrossRef]
41. Zhu, H.Y.; Shi, L.W.; Li, S.Q.; Zhang, S.B.; Xia, W.S. Pressure effects on structural, electronic, elastic and lattice dynamical properties of XSi_2 (X = Cr, Mo, W) from first principles. *Int. J. Mod. Phys. B* **2018**, *32*, 1850120. [CrossRef]
42. Yang, A.C.; Bao, L.K.; Peng, M.J.; Duan, Y.H. Explorations of elastic anisotropies and thermal properties of the hexagonal $TMSi_2$ (TM = Cr, Mo, W) silicides from first-principles calculations. *Mater. Today Commun.* **2021**, *27*, 102474. [CrossRef]
43. Zhu, H.Y.; Shi, L.W.; Li, S.Q.; Zhang, S.B.; Xia, W.S. Effects of biaxial strains on electronic and elastic properties of hexagonal XSi_2 (X = Cr, Mo, W) from first-principles. *Solid State Commun.* **2018**, *270*, 99–106. [CrossRef]

Disclaimer/Publisher's Note: The statements, opinions and data contained in all publications are solely those of the individual author(s) and contributor(s) and not of MDPI and/or the editor(s). MDPI and/or the editor(s) disclaim responsibility for any injury to people or property resulting from any ideas, methods, instructions or products referred to in the content.

Article

Investigation of the Quenching Sensitivity of the Mechanical and Corrosion Properties of 7475 Aluminum Alloy

Puli Cao [1], Guilan Xie [1], Chengbo Li [1,2,*], Daibo Zhu [1,*], Di Feng [3], Bo Xiao [1] and Cai Zhao [1]

1. School of Mechanical Engineering and Mechanics, Xiangtan University, Xiangtan 411105, China; xtucpl@126.com (P.C.); xieguilan@xtu.edu.cn (G.X.); m17369283823@163.com (B.X.); zcing0106@163.com (C.Z.)
2. Guangdong Xing Fa Aluminum Co., Ltd., Foshan 528137, China
3. School of Materials Science and Engineering, Jiangsu University of Science and Technology, Zhenjiang 212003, China; difeng1984@just.edu.cn
* Correspondence: xtulicb@xtu.edu.cn (C.L.); daibozhu@xtu.edu.cn (D.Z.); Tel.: +86-0731-5829-2209 (C.L.)

Abstract: Based on end-quenching experiments combined with conductivity, hardness testing, and microstructural characterization, the quenching sensitivity of the mechanical and corrosion properties of 7475 aluminum alloy was investigated. The study revealed that as the quenching rate decreased, both the mechanical properties and exfoliation corrosion resistance exhibited increased quenching sensitivity. With the quenching rate decreasing from 31.9 °C/s to 2.5 °C/s, the conductivity increased by 4.1%IACS, the hardness decreased by 31%, the exfoliation corrosion grade transitioned from EC to ED, and the maximum exfoliation corrosion depth increased from 237 µm to 508 µm. As the quenching rate decreased, the η phase sequentially precipitated at recrystallized grain boundaries (RGBs), E phase particles, and subgrain boundaries (SGBs), while the T phase primarily precipitated on E phase particles. Furthermore, the significant precipitation of η and T phases led to a notable reduction in the quantity of age-precipitated phases, an increase in their size, and poor coherency with the matrix, resulting in decreased mechanical properties and a higher quenching sensitivity of the mechanical performance. Meanwhile, with the reduction in quenching rate, the size and spacing of grain boundary precipitated phases increased, the Zn and Mg contents of grain boundary precipitated phases increased, and the Precipitation Free Zone (PFZ) widened, leading to decreased exfoliation corrosion resistance and higher quenching sensitivity of the exfoliation corrosion performance.

Keywords: high-strength aluminum alloy; mechanical; corrosion properties; quenching sensitivity; microstructure

Citation: Cao, P.; Xie, G.; Li, C.; Zhu, D.; Feng, D.; Xiao, B.; Zhao, C. Investigation of the Quenching Sensitivity of the Mechanical and Corrosion Properties of 7475 Aluminum Alloy. *Metals* **2023**, *13*, 1656. https://doi.org/10.3390/met13101656

Academic Editor: Frank Czerwinski

Received: 12 August 2023
Revised: 16 September 2023
Accepted: 25 September 2023
Published: 27 September 2023

Copyright: © 2023 by the authors. Licensee MDPI, Basel, Switzerland. This article is an open access article distributed under the terms and conditions of the Creative Commons Attribution (CC BY) license (https://creativecommons.org/licenses/by/4.0/).

1. Introduction

Al-Zn-Mg-Cu aluminum alloys, due to their high strength resulting from solid solution, quenching, and aging processes, are widely utilized as structural materials, being particularly extensively employed in the aerospace industry [1,2]. Quenching is a crucial step in the preparation of high-strength aluminum alloy materials. Rapid quenching leads to a high degree of supersaturated solid solution in the alloy, and subsequent aging results in the precipitation of numerous strengthening phases, thus achieving high strength. However, excessively high quenching rates often lead to elevated residual stresses [3]. Reducing the quenching rate can mitigate the residual stresses in the alloy, but this may result in decreased alloy performance. This phenomenon, where the alloy's performance diminishes after aging as the quenching rate decreases, is referred to as quenching sensitivity [4].

In general, for high-strength aluminum alloys, as the quenching rate decreases, the quantity and size of quenching-precipitated phases increase. This leads to a reduction in the solute atom concentration and vacancy concentration in the alloy after quenching. Subsequently, the number of precipitations strengthening phases decreases upon aging, resulting in a decline in mechanical properties and an increase in the quenching sensitivity of the mechanical performance [5–8]. Liu Shengdan et al. [9] found that in 7055 aluminum alloy, the quenching sensitivity of mechanical performance increases as the quenching rate decreases. Li Peiyue et al. [10] studied the quenching sensitivity of 7050 aluminum alloy using spray end-quenching and discovered that, as the quenching rate decreases, mechanical properties decrease, and quenching sensitivity increases. Liu et al. [11] found that the quenching sensitivity of the mechanical performance of 7085 alloy increases as the quenching rate decreases. Zheng Pengcheng [12] and Ma Zhimin et al. [13] found that the quenching sensitivity of the mechanical performance of 7136 aluminum alloy increases as the quenching rate decreases.

Variations in the quenching rate also impact the size, composition, spacing of grain boundary precipitated phases, and width of the Precipitation Free Zone (PFZ), thereby exerting complex effects on the alloy's corrosion resistance. However, there is still some controversy regarding its influence on the quenching sensitivity of localized corrosion performance. Many studies have indicated that as the quenching rate decreases, the exfoliation corrosion resistance of the alloy diminishes and the quenching sensitivity of the exfoliation corrosion performance increases. Marlaud et al. [14] found that the exfoliation corrosion resistance of 7449-T7651 alloy decreases with the decreasing quenching rate. Song et al. [15] observed an increase in the exfoliation corrosion sensitivity of the AA7050 alloy with a decreasing quenching rate. Li Dongfeng et al. [16] discovered that as the quenching rate decreased from 2160 °C/min to 100 °C/min, the exfoliation corrosion grade of Al-5Zn-3Mg-1Cu alloy sheets gradually shifted from P grade to ED grade. Liu et al. [17] revealed that the maximum exfoliation corrosion depth of AA7055 alloy gradually increases as the quenching rate decreases, indicating an ascending trend in exfoliation corrosion sensitivity. Ma et al. [13] also found an increase in the exfoliation corrosion sensitivity of 7136 aluminum alloy with a decreasing quenching rate.

In summary, it is evident that high-strength aluminum alloys exhibit not only quenching sensitivity in terms of mechanical performance, but also significant quenching sensitivity in their corrosion properties. Therefore, this study utilizes end-quenching experiments in conjunction with Transmission Electron Microscopy (TEM), High-Resolution Transmission Electron Microscopy (HRTEM), and Scanning Transmission Electron Microscopy (STEM) to systematically investigate the types, nucleation sites, sizes, and morphologies of precipitated phases under different quenching rates. The study also discusses the precipitation behavior of these quenching-induced phases and their influence on the quenching sensitivity of mechanical and exfoliation corrosion performance. This research aims to provide a better understanding of the quenching precipitation behavior and the mechanisms behind quenching sensitivity in high-strength aluminum alloys.

2. Experiment

The experimental material used was a hot-rolled thick plate provided by a certain company. The casting temperature was 700 °C, followed by a 24 h homogenization heat treatment at 465 °C. The hot rolling temperature was 390 °C, with a deformation amount of 90%. The chemical composition (wt%) of the plate is shown in Table 1. Cross-sectional samples with dimensions of 25 mm × 25 mm were cut from the surface of the hot-rolled plate with a length of 125 mm for subsequent solution heat treatment and end-quenching. The samples were heated to 470 °C and held in an air furnace (TPS, New Columbia, PA, USA) for 2 h. They were then transferred to an end-quenching device [18] and rapidly water-cooled by spraying water onto the groove end until reaching room temperature. The quenching water temperature was approximately 20 °C. The quenched samples were subsequently subjected to artificial aging in an oil bath at 120 °C for 24 h. After aging, half of the samples were polished using sandpaper and subjected to hardness testing. The test was conducted using three parallel samples. Hardness measurements were taken along the rolling direction at 5-mm intervals, starting from the water-cooled end. The hardness tests were conducted using an HV-10B Vickers hardness tester (Suzhou Nanguang Electronic Technology, Suzhou, China) with a load of 3 kg. For exfoliation corrosion testing, slices (2 mm thick) were cut from the aged samples. The test was conducted using three parallel samples. The testing was performed following the GB/T 22639-2008 standard [19]. The area-to-volume ratio of the solution was 25 cm²/L, and the testing temperature was maintained at (25 ± 2) °C. After 48 h of corrosion, the samples were evaluated according to the standard using an EXCO solution (4 mol/L NaCl + 0.5 mol/L KNO₃ + 0.1 mol/L HNO₃). Metallographic samples were prepared from different locations and observed for corrosion using an XJP-6A metallographic microscope (Suzhou Hongtai Instrument, Suzhou, China). After coarse grinding, fine grinding, and polishing, the metallographic samples were observed, and the corrosion depth was measured under the microscope. Additionally, samples of the same size were taken, and thermocouples were embedded at different distances (3, 13, 23, 53, 78, and 98 mm) from the water-cooled end. Cooling curves were recorded during the end-quenching process at these positions, and the average cooling rates were calculated in the temperature range of 185 to 415 °C [20]. The calculated average cooling rates for the respective positions were 31.9, 17.5, 8.4, 3.3, 2.9, and 2.5 °C/s, as shown in Figure 1. Thin slices (2 mm thick) were taken and subjected to water quenching at room temperature after solution treatment, resulting in a corresponding quenching rate of 960 °C/s.

Table 1. Chemical compositions of the alloy (wt%).

Zn	Mg	Cu	Cr	Fe	Si	Al
5.6	2.5	1.6	0.25	0.10	0.05	Bal

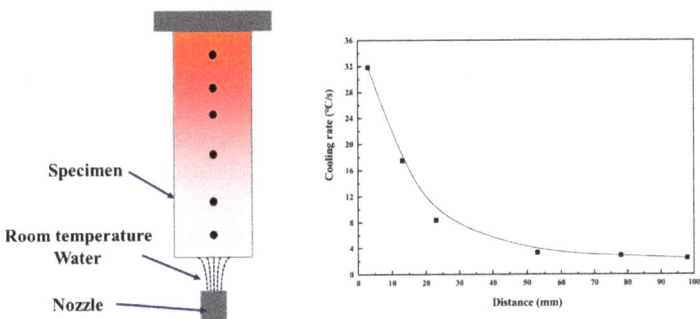

Figure 1. Schematic diagram of end quenching and quenching rate curve.

After aging, samples were extracted from the end-quenched specimens for microstructural analysis. The samples were first thinned by grinding to a thickness of approximately 60–80 μm. Circular discs with a diameter of 3 mm were then punched out. Thinning was further performed using dual-jet polishing in a solution containing 80% methanol and 20% nitric acid. The electrolyte temperature was controlled at around −25 °C using liquid nitrogen. Subsequently, the precipitated phases from the quenched samples were observed using a Tecnai G2 F20 (FEI, Eindhoven, The Netherlands) TEM and a Titan G2 60–300 (FEI, Eindhoven, The Netherlands) STEM. JMatPro 8.0 (Sente Software, London, UK) software was employed to calculate Time-Temperature-Transformation (TTT) curves and Continuous Cooling Transformation (CCT) curves.

3. Results

3.1. Electrical Conductivity and Hardness Tests

Figure 2 shows the conductivity curve. From the graph, it is evident that the alloy's conductivity increases with an increase in the distance from the water-cooled end. When the distance from the end is less than 60 mm, the conductivity rapidly increases with the distance, followed by a smaller increment in conductivity as the distance further increases. As the distance from 3 mm to 98 mm increases, the conductivity rises from 28.9 %IACS to

33 %IACS, resulting in a conductivity difference of 4.1 %IACS between the two ends. The conductivity in the quenched state serves as a reliable indicator of the quenching-induced precipitation behavior.

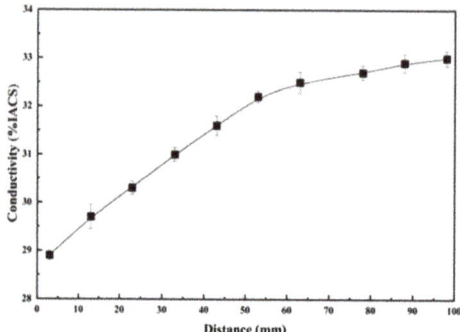

Figure 2. Conductivity curve.

Figure 3 represents the hardenability curve. From Figure 3a, it is evident that as the distance from the water-cooled end increases, the hardness gradually decreases. Within the region where the distance is less than 63 mm, the hardness value rapidly decreases with an increase in the distance, while beyond 63 mm, the change in hardness value with distance is relatively small. To further study the variations in hardness, based on the hardenability curve shown in Figure 3a, the retained hardness values were calculated, leading to the retained hardness curve depicted in Figure 3b. The trend of this curve is consistent with that of Figure 3a. Beyond a distance of 63 mm, the retained hardness value stabilizes around 70%, indicating a reduction in hardness of approximately 30%. At a distance of 98 mm, the hardness decreases by 31%.

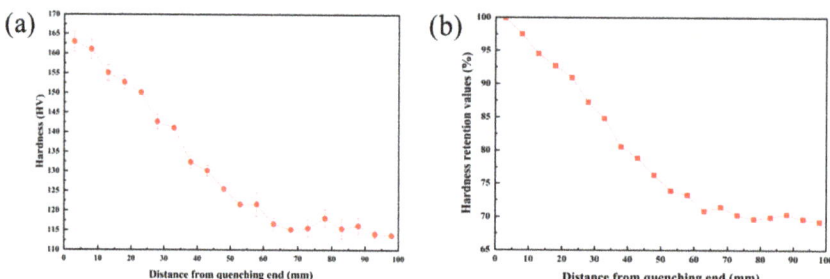

Figure 3. Hardenability curve (a) and hardness retention curve (b).

3.2. Exfoliation Corrosion

Figure 4 presents macroscopic images of the end-quenched samples immersed in EXCO solution for different durations. The regions farther from the water-cooled end exhibit a higher quantity of bubbles and more intense reactions. After a 2-h immersion, the sample surfaces show no significant corrosion. After 6 h of immersion, slight pitting corrosion is observed on the sample surface (Figure 4). In the areas beyond 23 mm from the water-cooled end, the corrosion grade is classified as PB. As the immersion time increases, the corrosion severity intensifies. After 12 h of immersion, significant corrosion is evident on the sample surface. Noticeable exfoliation is observed in the regions far from the water-cooled end, resulting in an EB corrosion grade. Extensive corrosion products are generated (Figure 4b). With prolonged immersion, severe delamination and exfoliation occur on the surface. After 24 h of immersion, exfoliation corrosion products become prominent, particularly in the regions far from the water-cooled end (Figure 4c). In areas with distances less than 23 mm, the corrosion grade is categorized as EA, while in regions beyond 23 mm, the grade is classified as EC. Following 48-h immersion, severe exfoliation corrosion is evident, with more corrosion products in positions farther from the water-cooled end. A substantial amount of corrosion products detaches, and the corrosion extends into the deeper metal interior. In regions beyond 23 mm, the corrosion grade is classified as ED (Figure 4d).

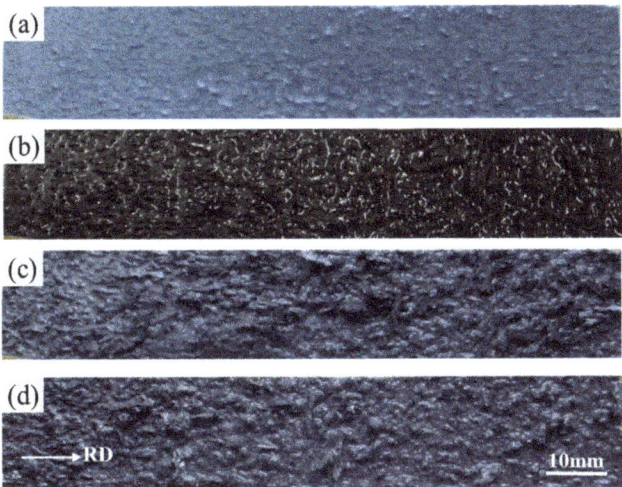

Figure 4. Corrosion morphology of end-quenched sample after soaking in EXCO solution for different times (spray end is on the left): (**a**) 6 h; (**b**) 12 h; (**c**) 24 h; (**d**) 48 h.

According to the GB/T 22639-2008 standard, the exfoliation corrosion degree of the samples was rated. It is evident from the graph that corrosion becomes increasingly severe with prolonged immersion. Exfoliation corrosion is more pronounced in regions with lower quenching rates than in those with higher quenching rates. Beyond 23 mm from the water-cooled end, the severity of exfoliation corrosion is notably high, with minimal difference in corrosion grades. After 48 h of immersion, samples with quenching rates above 8.4 °C/s are rated as EC. Samples with quenching rates below 8.4 °C/s are rated as ED.

Figure 5 displays cross-sectional metallographic images of the end-quenched samples after exfoliation corrosion at different quenching positions. It is evident from the images that the corrosion depth of the samples increases with a decrease in the quenching rate. Lower quenching rates lead to more pronounced exfoliation corrosion. When the quenching rate is 8.4 °C/s, the sample surfaces exhibit typical layered exfoliation corrosion morphology. The expansion of exfoliation corrosion products creates stress that lifts the metal layers one by one, resulting in severe exfoliation. The maximum and average exfoliation corrosion depths are shown in Figure 6b. From the graph, it can be observed that as the quenching rate decreases, both the maximum and average exfoliation corrosion depths increase. Moreover, lower quenching rates correspond to a greater increase in corrosion depth. At a quenching rate of 31.9 °C/s, the maximum and average exfoliation corrosion depths are 237 µm and 203 µm, respectively. At a quenching rate of 2.5 °C/s, the maximum and average exfoliation corrosion depths are 508 µm and 418 µm, respectively.

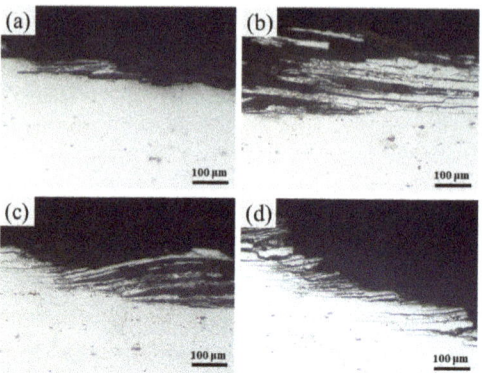

Figure 5. Metallographic photos of cross-section after spalling corrosion of samples at different quenching positions: (**a**) 31.9 °C/s, (**b**) 8.4 °C/s, (**c**) 3.3 °C/s, (**d**) 2.5 °C/s.

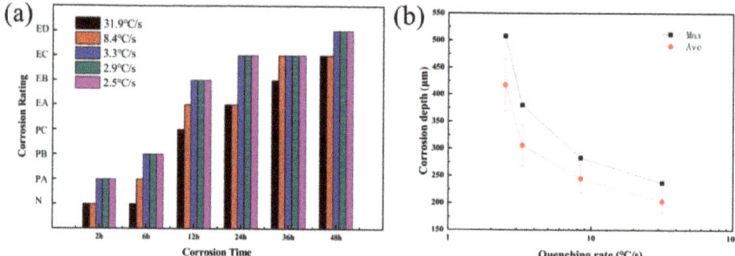

Figure 6. Spalling corrosion degree rating (**a**) and corrosion depth (**b**) of end-quenched sample.

3.3. Microstructure

Figure 7 presents TEM images of samples quenched at a rate of 960 °C/s. As observed in Figure 7a, a significant number of dispersed particles are present within the aluminum matrix. These particles exhibit irregular shapes and non-uniform sizes, ranging from circular and triangular to elongated forms, with dimensions ranging from 50 to 150 nm. Notably, these particles are non-coherent with the matrix [8]. Energy-dispersive X-ray spectroscopy (EDS) analysis results (Figure 7b) indicate that these particles consist of 81.5% Al, 2.4% Zn, 10.8% Mg, 1.8% Cu, and 3.5% Cr (atomic fractions). It is likely that the Cr-containing dispersed particles are associated with the E phase ($Al_{18}Cr_2Mg_3$).

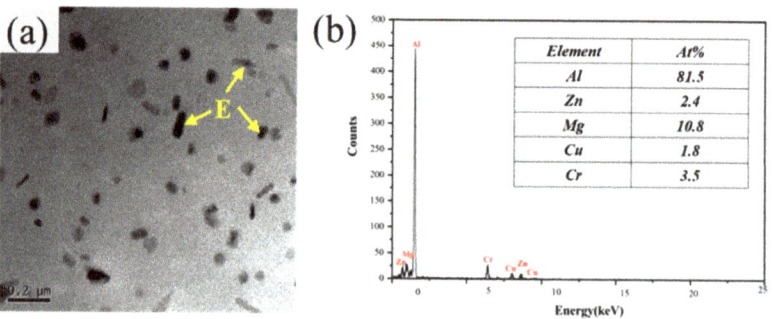

Figure 7. TEM photos at a quenching rate of 960 °C/s: (**a**) Bright field phase, (**b**) EDS.

Figure 8 displays TEM images of samples quenched at a rate of 31.9 °C/s. In Figure 8a, it can be observed that quenching precipitation occurs at recrystallized grain boundaries (RGBs), while no quenching precipitation is observed at subgrains (SGs) and their boundaries. Figure 8b reveals that the intragranular η phase nucleates and precipitates on E phase particles, with η phase sizes ranging from 100 to 150 nm. Simultaneously, age-related precipitates can be observed within the grain. Figure 8c demonstrates the fine and uniform dispersion of age-related precipitates within the grain's matrix. These precipitates exhibit spherical and rod-like morphologies. For a more detailed examination of the age-related precipitates, high-resolution TEM was employed to observe from the <011> direction. The HRTEM images (Figure 8d,e) reveal that the size of the age-related η' phases is 5–10 nm, indicating a relatively good coherence with the matrix. At this stage, the age-strengthening effect is notable, resulting in higher hardness.

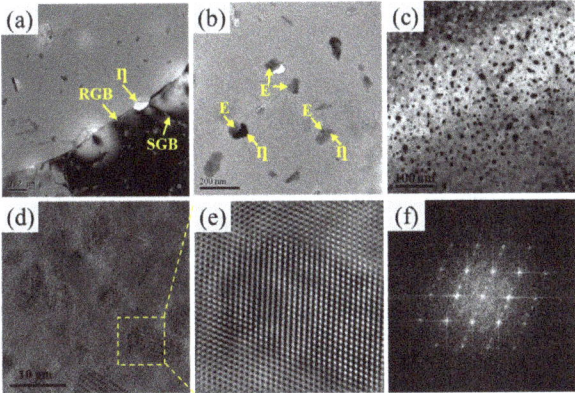

Figure 8. TEM photo of a quenching rate of 31.9 °C/s: (**a**) grain boundary, (**b**) intragranular, (**c**) aging precipitate, (**d**) <011>$_{Al}$ HRTEM, (**e**) Inverse Fast Fourier Transformation (IFFT), (**f**) Fast Fourier Transformation (FFT).

Figure 9 displays HAADF-STEM images of samples quenched at a rate of 8.4 °C/s. In Figure 9a, it is evident that prominent quenching precipitates are present at RGBs, while SGs and their boundaries also exhibit quenching precipitates. Within the matrix, a substantial amount of quenching precipitates, identified as the η phase, can be observed. These precipitates exhibit plate-like shapes of uneven sizes, with some reaching lengths of up to 400 nm. Additionally, hexagonal T-phase precipitates with dimensions ranging from 100 to 200 nm are present within the matrix. Furthermore, numerous smaller quenching precipitates are distributed within the matrix, as shown in Figure 9b. Based on EDS analysis, it is revealed that the η phase primarily contains 75.7% Al, 11.5% Zn, 9.6% Mg, and 3.2% Cu (at%). The T phase mainly comprises 86.1% Al, 5.8% Zn, 6.3% Mg, and 1.8% Cu (at%).

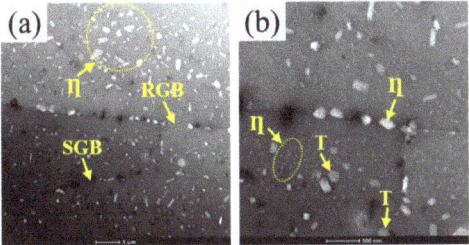

Figure 9. HAADF-STEM photos with a quenching rate of 8.4 °C/s: (**a**) low power, (**b**) high power.

Figure 10 presents TEM images of samples quenched at a rate of 3.3 °C/s. In Figure 10a, a significant amount of quenching precipitates, identified as the η phase, can be observed within the grains. These precipitates exhibit plate-like shapes of considerable size, with some reaching lengths of up to 500 nm. Additionally, hexagonal T phase precipitates are also present. In Figure 10b, it can be observed that a substantial quantity of η phase quenching precipitates exists within SGs and at their boundaries. The sizes of these quenching precipitates vary, with some being larger and others being much smaller.

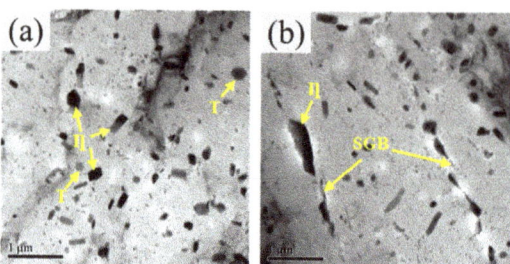

Figure 10. TEM photos at a quenching rate of 3.3 °C/s. (**a**) intracrystalline, (**b**) grain boundary.

Figure 11 depicts TEM images of samples quenched at a rate of 2.5 °C/s. In Figure 11a, an abundance of η and T phases are observed within the grains. Simultaneously, numerous η phase precipitates are observed within SGs and at their boundaries, displaying uneven sizes and being distributed along the deformation direction, as shown in Figure 11b. Observations at higher magnification in Figure 11c reveal age-related precipitates. It is evident that a noticeable Precipitation-Free Zone (PFZ) forms around a wider area of the quenching precipitates, where age-related precipitates are absent. Age-related precipitates near the quenching precipitates and grain boundaries are relatively larger and less abundant. HRTEM observations from the <011> direction (Figure 11d) indicate that the size of the age-related η' phase is 10–20 nm, indicating relatively poor coherence with the matrix. At this stage, the age-strengthening effect is relatively weak, resulting in lower hardness and a significant hardness drop of 31% in this region.

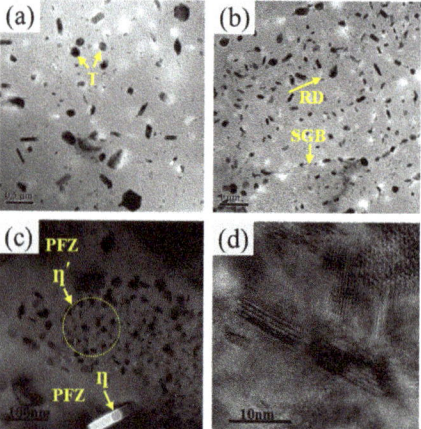

Figure 11. TEM photo of quenching rate of 2.5 °C/s: (**a**) intragranular, (**b**) grain boundary, (**c**) aging precipitate, (**d**) <011>$_{Al}$ HRTEM.

Figure 12 illustrates the HAADF-STEM images of samples quenched at a rate of 2.5 °C/s. In Figure 12a, large-sized quenched precipitates are observed at RGBs, with sizes around 500 nm. Some of these quenched precipitates have been corroded, leaving behind dark features, indicating that the precipitates at grain boundaries are susceptible to corrosion. Precipitates are also observed at subgrain boundaries (SGBs), with sizes around 250 nm. Within the grains, in addition to the larger-sized quenched precipitates, a multitude of smaller-sized quenched precipitates are observed. From the higher magnification image in Figure 12b, it is evident that quenched precipitates of η phase are precipitating on E-phase particles, resulting in larger and less abundant age-related precipitates around them. A distinct Precipitation-Free Zone (PFZ) is observed surrounding the η phase, leading to a significant reduction in mechanical performance.

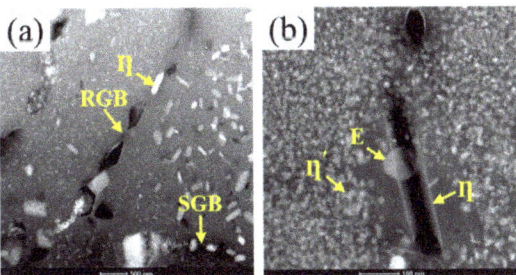

Figure 12. HAADF-STEM photos with a quenching rate of 2.5 °C/s: (**a**) grain boundary, (**b**) intragrain.

Figure 13 depicts the composition of grain boundary precipitates at different quenching rates. As shown in the figure, with decreasing quenching rates, the content of Zn, Mg, and Cu in grain boundary precipitates increases. Zn content exhibits the most rapid increase, followed by Mg, while the increase in Cu content is comparatively slower. At a quenching rate of 31.9 °C/s, the grain boundary precipitates contain lower levels of Zn, Mg, and Cu. In contrast, at a quenching rate of 2.5 °C/s, both Zn and Mg content show significant increments, and there is also a notable increase in Cu content.

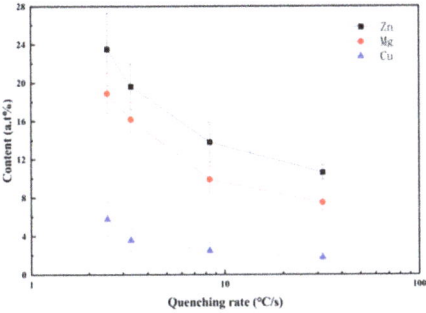

Figure 13. Components of grain boundary precipitated phase at different quenching rates.

3.4. TTT and CCT Curves

Figure 14 illustrates the TTT and CCT curves. In Figure 14a, the curve for the η phase is on the left side, while the curve for the T phase is on the right side. The nose temperatures for the η phase and T phase are 332.1 and 301.2 °C, respectively, corresponding to transformation times of 20.6 and 47.1 s, respectively. From the CCT curve in Figure 14b, it can be observed that during the decomposition of the supersaturated solid solution, the η phase precipitates first, followed by the T phase.

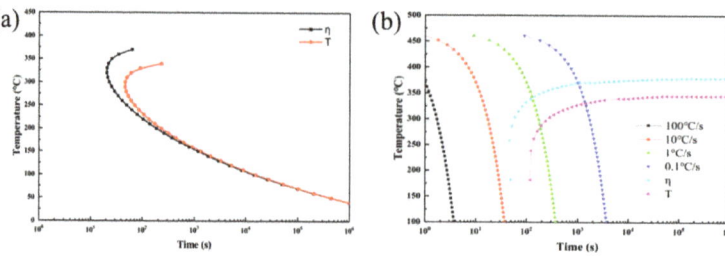

Figure 14. (a) TTT, (b) CCT curve.

4. Discussion

The quenched precipitates and their nucleation sites at different quenching rates are summarized in Table 2. It is evident from the table that as the quenching rate decreases, both the types of quenched precipitates and their nucleation sites increase. The quenched precipitates include the η phase and the T phase. The η phase preferentially precipitates at RGBs s, while it also precipitates on the E phase particles as well as at SGBs. Combined with Figure 14, it can be inferred that the η phase precipitates first during the quenching process, followed by the T phase.

Table 2. Quenching precipitates and nucleation positions at different quenching rates.

Quenching Rate (°C/s)	RGBs	E Particle	SGBs
960	—	—	—
31.9	η	η	—
8.4	η	η, T	η
3.3	η	η, T	η
2.5	η	η, T	η

The sizes of the η phase precipitates at different quenching rates are presented in Table 3. As indicated in the table, the size of the η phase precipitates gradually increases with the decreasing quenching rate. The largest η phase precipitates are observed at RGBs, followed by those on the E phase particles, while the η phase precipitates at SGBs are smaller in size.

Table 3. Size of the η phase at different quenching rates.

Quenching Rate (°C/s)	RGBs	E Particle	SGBs
960	—	—	—
31.9	151.1 ± 55.9	128.8 ± 48.7	—
8.4	235.7 ± 88.2	185.6 ± 78.1	119.4 ± 57.3
3.3	345.2 ± 102.6	240.7 ± 90.5	167.7 ± 69.4
2.5	526.1 ± 192.5	369.8 ± 138.6	251.4 ± 102.5

During the quenching cooling process, the quenched precipitation phase at the grain boundaries is primarily the η phase, which mainly contains Zn, Mg, and Cu elements. The diffusion rates of Zn, Mg, and Cu in the Al solid solution are in the order Zn > Mg > Cu [21,22]. Generally, the nucleation activation energy of the η phase is not very high. At room temperature, the diffusion rate of Zn in the Al matrix is 1.64×10^{-12} cm^2/s, while that of Mg is 7.33×10^{-15} cm^2/s. The activation energy for aluminum self-diffusion is 142.8 kJ/mol, for Mg in aluminum it is 135.1 kJ/mol, and for Zn in aluminum it is 106.1 kJ/mol. Therefore, it can be inferred that with increasing temperature, the diffusion of Zn atoms in the system becomes relatively easier. The η phase preferentially nucleates at the RGBs with high interfacial energy. Since the grain boundaries serve as fast diffusion paths for solute atoms such as Zn and Mg, the η phase preferentially nucleates and grows at the grain boundaries [23,24]. Subsequently, nucleation occurs on E-phase particles, and finally at the SGBs. During the alloy quenching process, a significant amount of η phase preferentially precipitates, consuming a substantial number of Zn atoms. This leads to a low local concentration of Zn atoms. Meanwhile, the high interfacial energy of the E phase particles provides a favorable site for the nucleation of the T phase, resulting in the precipitation of the T phase (Figure 9).

During quenching, precipitation occurs due to desaturation, leading to a reduction in the supersaturation of the solid solution and a weakening of the lattice distortion, thereby causing an increase in conductivity. Locations closer to the water-cooled end experience a higher quenching cooling rate, resulting in the formation of a highly supersaturated solid solution and significant lattice distortion, which leads to lower conductivity. In contrast, positions farther from the water-cooled end experience a lower quenching cooling rate, causing a significant amount of equilibrium phase to desaturate, resulting in a lower supersaturation of the solid solution and reduced lattice distortion, hence leading to higher conductivity.

As the quenching rate decreases, quenching-induced phases precipitate at both grain boundaries and within the grains, with their sizes gradually increasing. The precipitation of η and T phases consumes the Zn and Mg solute atoms in the matrix, resulting in a reduction in solute atom concentration and vacancy concentration in the alloy after quenching. Subsequently, during aging, the precipitation-strengthening phases exhibit reduced quantities and increased sizes. These phases exhibit relatively poor lattice compatibility with the matrix (Figure 11c,d), leading to weaker aging strengthening effects. Furthermore, the quenched precipitation phases at the grain boundaries and within the grains can continue to grow by absorbing surrounding solute atoms during the aging process, forming broader PFZs (Figure 12b). The broadened PFZs further contribute to a reduction in mechanical properties. Consequently, the quenching sensitivity of the mechanical properties increases as the quenching rate decreases.

As the quenching rate decreases, the alloy's resistance to exfoliation corrosion diminishes. Considering the observed microstructural features (Figures 8a, 9, 12a and 13), this phenomenon is mainly attributed to the increased size and spacing of grain boundary precipitates, elevated Zn and Mg content in the grain boundary precipitates, and broadening of the PFZs with decreasing quenching rates. Prior investigations have indicated that the potential of the grain boundary η phase is -1.05 V, PFZ is -0.85 V, and the potential of the matrix within grains is -0.75 V [25]. In corrosive environments, the differing corrosion potentials of various phases in the grain boundary regions of the matrix solid solution can lead to galvanic corrosion, resulting in along-grain boundary corrosion in high-strength aluminum alloys. The formation of corrosion microcells between grain boundary precipitates (η phase), PFZs, and phases within the grain can establish anodic dissolution sites. The grain boundary η phase acts as the anodic phase and is preferentially dissolved, and PFZs often function as anodic sites that are susceptible to corrosion, thereby creating pathways for along-grain boundary anodic dissolution channels [26–28]. As the quenching rate decreases, coarser η phase precipitates form at grain boundaries, exhibiting increased and non-uniform sizes, along with elevated Zn and Mg content in the precipitates. Simultaneously, PFZs broaden, further promoting grain boundary dissolution. This corrosion process generates substantial corrosion products during corrosion, leading to exfoliation corrosion and ultimately reducing the alloy's resistance to exfoliation corrosion.

5. Conclusions

(1) As the quenching rate decreases, the electrical conductivity increases, and the quenching sensitivity of both the mechanical properties and exfoliation corrosion increases. With a decrease in the quenching rate from 31.9 °C/s to 2.5 °C/s, the electrical conductivity increases by 4.1 %IACS, hardness decreases by 31%, and the exfoliation corrosion grade transitions from EC to ED, with the maximum exfoliation corrosion depth increasing from 237 μm to 508 μm.

(2) With a reduction in the quenching rate, the variety, nucleation sites, size, and quantity of quenching-induced precipitates increase. During quenching, η phase precipitates first, followed by the precipitation of T phase. η phase predominantly nucleates at RGBs, on E phase particles, and at SGBs, while T phase primarily forms on E phase particles.

(3) Decreasing quenching rates lead to abundant precipitation of η and T phases within the matrix and along grain boundaries. This results in a significant reduction in the quantity and an increase in the size of precipitates during aging. These precipitates also exhibit poorer coherency with the matrix, and a PFZ forms, contributing to reduced mechanical properties and heightened sensitivity to quenching-induced variations. Moreover, lower quenching rates lead to larger precipitate sizes and spacings at grain boundaries, increased Zn and Mg content in grain boundary precipitates, and broadened PFZs, all contributing to diminished resistance against exfoliation corrosion and heightened sensitivity to exfoliation corrosion performance.

Author Contributions: Investigation and Writing—original draft, P.C.; Software and Writing—original draft, G.X.; Conceptualization, and Methodology, C.L.; Supervision and Funding Acquisition, D.Z.; Validation, Data curation and Formal analysis, D.F., B.X. and C.Z. All authors have read and agreed to the published version of the manuscript.

Funding: This study was supported by the National Natural Science Foundation of China (52205421), Guangxi Science and Technology Major Project (AA23023028), the Key laboratory open project of Guangdong Province (XF20230330-XT), the school-enterprise, industry-university-research cooperation project (2023XF-FW-32), the science and technology innovation Program of Hunan Province, China (2021RC2087, and 2022JJ30570), and the Key Research and Development Program of Zhenjiang City (GY2021003 and GY2021020).

Institutional Review Board Statement: Not applicable.

Informed Consent Statement: Not applicable.

Data Availability Statement: Data will be made available on request.

Conflicts of Interest: The authors declare no conflict of interest.

References

1. Deng, Y.L.; Zhang, X.M. Development of aluminium and aluminium alloy. *Chin. J. Nonferrous Met.* **2019**, *29*, 2115–2141.
2. Zhou, B.; Liu, B.; Zhang, S. The advancement of 7XXX series aluminum alloys for aircraft structures: A review. *Metals* **2021**, *718*, 718. [CrossRef]
3. Lin, G.Y.; Zheng, X.Y.; Feng, D.; Yang, W.; Peng, D.S. Research development of quenching-induced residual stress of aluminum thick plates. *Mater. Rev.* **2008**, *22*, 70–74.
4. Jiang, F.Q.; Huang, J.W.; Jiang, Y.G.; Xu, C.L. Effects of quenching rate and over-aging on microstructures, mechanical properties and corrosion resistance of an Al-Zn-Mg (7046A) alloy. *J. Alloys Compd.* **2021**, *854*, 157272. [CrossRef]
5. Pan, T.A.; Tzeng, Y.C.; Bor, H.Y.; Liu, K.H.; Lee, S.L. Effects of the coherency of Al_3Zr on the microstructures and quench sensitivity of Al-Zn-Mg-Cu alloys. *Mater. Today Commun.* **2021**, *28*, 102611. [CrossRef]
6. Chen, J.S.; Li, X.W.; Xiong, B.Q.; Zhang, Y.A.; Li, Z.H.; Yan, H.W.; Liu, H.W.; Huang, S.H. Quench sensitivity of novel Al-Zn-Mg-Cu alloys containing different Cu contents. *Rare Met.* **2020**, *39*, 1395–1401. [CrossRef]
7. Peng, Y.H.; Liu, C.Y.; Wei, L.L.; Jiang, H.J.; Ge, Z.J. Quench sensitivity and microstructures of high-Zn content Al-Zn-Mg-Cu alloys with different Cu contents and Sc addition. *Trans. Nonferrous Met. Soc. China* **2021**, *31*, 24–35. [CrossRef]
8. Nie, B.H.; Liu, P.Y.; Zhou, T.T. Effect of compositions on the quenching sensitivity of 7050 and 7085 alloys. *Mater. Sci. Eng. A* **2016**, *667*, 106–114. [CrossRef]
9. Liu, S.D.; Li, C.B.; Han, S.Q.; Deng, Y.L.; Zhang, X.M. Effect of natural aging on quench-induced inhomogeneity of microstructure and hardness in high strength 7055 aluminum alloy. *J. Alloys Compd.* **2015**, *625*, 34–43. [CrossRef]
10. Li, P.Y.; Xiong, B.Q.; Zhang, Y.A.; Li, Z.H.; Zhu, B.H.; Wang, F.; Liu, H.W. Effect of quench behavior of 7050 Al alloy. *Chin. J. Nonferrous Met.* **2011**, *21*, 961–967.
11. Liu, S.D.; Li, Q.; Lin, H.Q.; Sun, L.; Long, T.; Ye, L.Y.; Deng, Y.L. Effect of quench-induced precipitation on microstructure and mechanical properties of 7085 aluminum alloy. *Mater. Des.* **2017**, *132*, 119–128. [CrossRef]
12. Zheng, P.C.; Ma, Z.M.; Liu, H.L.; Liu, S.D.; Xiao, Y.; Wei, W.C. Effect of quenching rate on microstructure and mechanical properties of 7136 aluminum alloy. *Chin. J. Nonferrous Met.* **2021**, *31*, 2348–2356.
13. Ma, Z.M.; Jia, L.; Yang, Z.S.; Liu, S.D.; Zhang, Y. Effect of cooling rate and grain structure on the exfoliation corrosion susceptibility of AA7136 alloy. *Mater. Charact.* **2020**, *168*, 110533. [CrossRef]

14. Marlaud, T.; Malki, B.; Henon, C.; Deschamps, A.; Baroux, B. Relationship between alloy composition, microstructure and exfoliation corrosion in Al-Zn-Mg-Cu alloys. *Corros. Sci.* **2011**, *53*, 3139–3149. [CrossRef]
15. Song, F.X.; Zhang, X.M.; Liu, S.D.; Tan, Q.; Li, D.F. The effect of quench rate and over-ageing temper on the corrosion behaviour of AA7050. *Corros. Sci.* **2014**, *78*, 276–286. [CrossRef]
16. Li, D.F.; Zhang, X.M.; Liu, S.D.; Yin, B.W.; Lei, Y. Effect of quenching rate on exfoliation corrosion of Al-5Zn-3Mg-1Cu aluminum alloy thick plate. *Hunan Univ. J. Nat. Sci.* **2015**, *42*, 47–52.
17. Liu, S.D.; Chen, B.; Li, C.B.; Dai, Y.; Deng, Y.L.; Zhang, X.M. Mechanism of low exfoliation corrosion resistance due to slow quenching in high strength aluminium alloy. *Corros. Sci.* **2015**, *91*, 203–212. [CrossRef]
18. Zhang, X.M.; Deng, Y.L.; You, J.H.; Zhang, Y.; Liu, X.H.; He, J.T.; Zhou, Z.P. Measurement Device and Method of Quenched Depth for Aluminum Alloy. CN101013124, 8 August 2007.
19. GB/T 22639-2008; Flake Corrosion Test Method for Aluminum Alloy Processed Products. General Administration of Quality Supervision, In-spection and Quarantine of the People's Republic of China and the Standardization Administration of the People's Republic of China. Standardization Administration of the People's: Beijing, China, 2008.
20. Liu, S.D.; Zhong, Q.M.; Zhang, Y.; Liu, W.J.; Zhang, X.M.; Deng, Y.L. Investigation of quench sensitivity of high strength Al-Zn-Mg-Cu alloys by time-temperature-properties diagrams. *Mater. Des.* **2010**, *31*, 3116–3120. [CrossRef]
21. Du, Y.; Chang, Y.A.; Huang, B.Y.; Gong, W.P.; Jin, Z.P.; Xu, H.H.; Yuan, Z.H.; Liu, Y.; He, Y.H.; Xie, F.Y. Diffusion coefficients of some solutes in fcc and liquid Al: Critical evaluation and correlation. *Mater. Sci. Eng. A* **2003**, *363*, 140–151. [CrossRef]
22. Beerwald, A. Diffusion of various metals in aluminum. *Z. Elektrochem. Angew. Phys. Chem.* **1939**, *45*, 789–795.
23. Zhang, X.M.; Liu, W.J.; Liu, S.D. Effect of processing parameters on quench sensitivity of an AA7050 sheet. *Mater. Sci. Eng. A* **2011**, *528*, 795–802. [CrossRef]
24. Liu, S.D.; Liu, W.J.; Zhang, Y. Effect of microstructure on the quench sensitivity of Al-Zn-Mg-Cu alloys. *J. Alloys Compd.* **2010**, *507*, 53–61. [CrossRef]
25. Song, F.X. *Investigation on Susceptibility to the Localized Corrosion of Aluminium 7050 Thick Plate*; Central South University: Changsha, China, 2014.
26. Chen, M.Y.; Xu, Z.; He, K.Z.; Liu, S.D.; Yong, Z. Local corrosion mechanism of an Al-Zn-Mg-Cu alloy in oxygenated chloride solution: Cathode activity of quenching-induced η precipitates. *Corros. Sci.* **2021**, *191*, 109743. [CrossRef]
27. Song, F.X.; Zhang, X.M. The effect of quench transfer time on microstructure and localized corrosion behavior of 7050-T6 Al alloy. *Mater. Corros.* **2014**, *65*, 1007–1016. [CrossRef]
28. Peng, X.; Chen, S.Y.; Chen, K.H.; Jiao, H.B.; Huang, L.P.; Zhang, Z.; Yang, Z. Enhancing the stress corrosion cracking resistance of a low-Cu containing Al-Zn-Mg-Cu aluminum alloy by step-quench and aging heat treatment. *Corros. Sci.* **2019**, *161*, 108184.

Disclaimer/Publisher's Note: The statements, opinions and data contained in all publications are solely those of the individual author(s) and contributor(s) and not of MDPI and/or the editor(s). MDPI and/or the editor(s) disclaim responsibility for any injury to people or property resulting from any ideas, methods, instructions or products referred to in the content.

Editorial

Aluminum Alloys and Aluminum-Based Matrix Composites

Di Feng [1,*], Qianhao Zang [1], Ying Liu [2] and Yunsoo Lee [3]

1. School of Materials Science and Engineering, Jiangsu University of Science and Technology, Zhenjiang 212100, China; qhzang@just.edu.cn
2. Department of Materials Science and Engineering, Nanjing University of Science and Technology, Nanjing 210094, China; liuying517@njust.edu.cn
3. Metallic Materials Division, Korea Institute of Materials Science (KIMS), Changwon 51508, Republic of Korea; yslee@kims.re.kr
* Correspondence: difeng1984@just.edu.cn; Tel.: +86-0511-84401184

1. Introduction and Scope

Due to air pollution and energy shortages in the contemporary world, weight lighting for transportation vehicles and energy conservation, as well as emission reductions, are necessary to achieve carbon neutrality and fuel conservation. As a structural material with high specific strength, good process performance, and abundant reserves, aluminum alloy is undoubtedly becoming a substitute for steel materials.

Since the invention of electrolytic aluminum technology, aluminum alloys have been widely used in the fields of aviation and automobiles. The aerospace industry mainly develops aluminum alloys with high strength, high toughness, and excellent stress corrosion resistance to meet the strict usage conditions. 2xxx series and 7xxx series aluminum alloys are typical structural materials for aviation [1]. The current research hotspot lies in the optimization of processing technology and improving the material composition. Powder metallurgy and spray deposition are typical innovative technologies that can avoid compositional segregation and obtain higher element contents. Research on aluminum matrix composites and superplastic aluminum alloy materials is also ongoing. In the industries of new energy vehicles and intelligent connected vehicles, 4xxx and 6xxx series aluminum alloys are widely used. The application of aluminum alloys in a vehicle body and chassis can reduce the weight of the entire vehicle by 20–40%, which effectively extends its range. For car wheels with a complex shape, high-pressure die-casting technology is now commonly used. Compared to steel wheels, the weight of Al-Si series wheels is greatly reduced, effectively reducing the vehicle's fuel consumption and carbon dioxide emissions. Automotive power battery shells made of aluminum alloy can reduce the weight by nearly 20%. The front and rear anti-collision beams made of optimized 7xxx aluminum alloy profiles have energy absorption values of no less than those of steel, achieving weight reduction and improving safety. In addition, aluminum alloys can also be used to manufacture components such as cylinder blocks, cylinder heads, crankshafts, connecting rods, and pistons for automotive engines. Plate and profile goods are commonly used products of aluminum alloys. Different deformation technologies are required to obtain the various shapes. The hot deformation behavior or the thermal deformation constitutive equation is a useful tool to obtain the optimized process parameters. Based on these mathematical models, the deformation defects, and even the service performances, of certain aluminum alloys can be predicted. Computational materials science can greatly shorten the cycle of material preparation processes. This Special Issue's scope embraces several types of aluminum alloys and the interdisciplinary work aimed at introducing the emerging area of technologies and theories.

Citation: Feng, D.; Zang, Q.; Liu, Y.; Lee, Y. Aluminum Alloys and Aluminum-Based Matrix Composites. *Metals* **2023**, *13*, 1870. https://doi.org/10.3390/met13111870

Received: 30 October 2023
Accepted: 3 November 2023
Published: 10 November 2023

Copyright: © 2023 by the authors. Licensee MDPI, Basel, Switzerland. This article is an open access article distributed under the terms and conditions of the Creative Commons Attribution (CC BY) license (https://creativecommons.org/licenses/by/4.0/).

2. Contributions

Ten articles have been published in the current Special Issue of *Metals*, encompassing the fields of hot deformation behavior, constitutive modeling, performance prediction modeling, structure designation, composition designation and quenching sensitivity. Current Special Issue papers can also be classified based on material composition, including AlZnMgCu, AlMg, AlSi, and AlLi alloys.

2.1. Hot Deformation Behavior

The stress–strain curve helps scholars understand the dynamic hardening and dynamic softening behaviors during hot deformation, such as dynamic recovery and dynamic recrystallization [2], which are the basis for optimizing hot-working process parameters. In addition, the constitutive models can also be established based on stress–strain curves under different temperatures and strain rates. The constitutive equation is a necessary model for finite element simulation of plastic deformation to obtain the deformation state, temperature distribution, and the stress concentration during processing. This Special Issue presents two typical constitutive models, which are often used to describe warm forging and hot compression, respectively.

2.2. Performance Prediction Modeling

The prediction of processing defects and performance is an important step during component production. An accurate damage model or state equation under a complex loading environment can precisely calculate the service life of components, laying a theoretical foundation for material selection and structural designation [3], effectively avoiding material failure and shortening the product development cycle. The Special Issue also presents two prediction models, which are applied to the real-time monitor of surface defects and the performance life under a cyclic tension–compression condition, respectively. The first-principle investigation for the mechanical properties of an Al-Mg-Zr alloy under uniaxial tension is also presented.

2.3. Composition Designation

The Zn/Mg ratio and Mg/Si ratio determine the strength of 7xxx and 6xxx aluminum alloys [4], respectively. The increase in Cu content is conducive to improvements in the aging hardening rate and the corrosion resistance. The element Cu is also one of the constituent elements of the nano strengthening phase in AlCu and AlCuLi alloys. For high-strength aluminum alloys, Fe and Si are impurities that can cause a sharp decrease in plasticity. However, Si is the main alloying element in a 4xxx alloy. The increase in Si content improves the fluidity of a AlSi alloy. Adding Cu and Mg to AlSi can also form the θ, β or Q phases, similar to those in AlCu and AlMgSi alloys [5]. Of course, the plasticity will be deteriorated when Si exceeds the eutectic composition. At this point, it is necessary to combine this with rapid solidification technology to improve the morphology of primary Si. It should be pointed out that there is a suitable content for any alloying element. Too little addition does not have a strengthening effect, while too much addition may lead to precipitation of the coarse second phase at the interface, reducing the strength, plasticity, and even corrosion resistance.

2.4. Quenching Sensitivity

Quenching is one heat treatment technology that obtains excellent strength, toughness and corrosion resistance, and so on. For high-performance aluminum alloys with a high alloying element content, a lower quenching cooling rate will cause a large number of alloying elements to precipitate along grain boundaries during the cooling, forming coarse and incoherent compounds [6]. Quenching precipitation greatly reduces the mechanical properties of the alloy while also deteriorating the corrosion resistance. Where are these coarse compounds formed? What are the ingredients? How does it affect the material properties? At what quenching rate level can cooling precipitation be suppressed? The answers to the above questions form the foundation for obtaining a high-quality supersaturated solid solution, which results a high-aging strengthening effect. Quenching sensitivity is particularly prominent in AlZnMgCu alloys. The 7085 aluminum alloy is currently known as the most excellent hardenability aluminum alloy.

2.5. Structure Designation

Even with the same material, different structural designations can be used to achieve different performances, such as increasing stiffness and improving a material's resistance to external loads [7]. In this research neighborhood, computational materials science is also an important application technology that can help researchers quickly judge the rationality of structures and reduce the number of physical experiments.

2.6. External Field-Assisted Manufacturing

External field-assisted manufacturing is a highly innovative technology that can be used to compensate for the shortcomings of traditional techniques, thereby obtaining a better processing experience or a superior performance. Lasers, magnetic fields, and ultrasound are commonly used external media [8]. The input of these external energy fields changes the processing rate and may also alter the law of the microstructure evolution, resulting in unexpected performances.

3. Conclusions and Outlook

With the vigorous development of the manned aerospace and intelligent driving vehicle industries, aluminum alloys have shown increasingly vigorous vitality. Whether it is the breakthrough in composition design concepts or the endless emergence of new technologies and processes, something has greatly expanded the application breadth and depth of aluminum alloys. This Special Issue gathers multiple related topics and provides an overview of the latest developments in aluminum alloys, with different compositions and their related technologies.

As Guest Editors of this Special Issue, we hope that these published papers can be helpful to scientists and engineers engaged in the research and development of high-performance aluminum alloys. We also hope that these contributors can establish connections, accomplish interdisciplinary and professional complementarity work, and then achieve greater successes. At the same time, we would like to warmly thank all the authors for their contributions, and all of the reviewers for their efforts in ensuring a high-quality publication. We offer our sincere thanks to the Editors of *Metals* for their continuous help and support during the preparation of this issue. In particular, my sincere thanks goes to Toliver Guo for his help and support.

Conflicts of Interest: The authors declare no conflict of interest.

List of Contributions

1. Feng, D.; Xu, R.; Li, J.; Huang, W.; Wang, J.; Liu, Y.; Zhao, L.; Li, C.; Zhang, H. Microstructure Evolution Behavior of Spray-Deposited 7055 Aluminum Alloy during Hot Deformation. *Metals* 2022, *12*, 1982; https://doi.org/10.3390/met12111982.
2. Tao, Y.; Wang, Y.; He, Q.; Xu, D.; Li, L. Comparative Study and Multi-Objective Crashworthiness Optimization Design of Foam and Honeycomb-Filled Novel Aluminum Thin-Walled Tubes. *Metals* 2022, *12*, 2163; https://doi.org/10.3390/met12122163.
3. Zhou, P.; Wang, D.; Nagaumi, H.; Wang, R.; Zhang, X.; Li, X.; Zhang, H.; Zhang, B. Microstructural Evolution and Mechanical Properties of Al-Si-Mg-Cu Cast Alloys with Different Cu Contents. *Metals* 2023, *13*, 98; https://doi.org/10.3390/met13010098.
4. Chen, G.; Zhao, C.; Shi, H.; Zhu, Q.; Shen, G.; Liu, Z.; Wang, C.; Chen, D. Research on the 2A11 Aluminum Alloy Sheet Cyclic Tension&Compression Test and Its Application in a Mixed Hardening Model. *Metals* 2023, *13*, 229; https://doi.org/10.3390/met13020229.

5. Tang, J.; Liu, S.; Zhao, D.; Tang, L.; Zou, W.; Zheng, B. An Algorithm for Real-Time Aluminum Profile Surface Defects Detection Based on Lightweight Network Structure. *Metals* 2023, *13*, 507; https://doi.org/10.3390/met13030507.
6. Xia, J.; Liu, R.; Zhao, J.; Guan, Y.; Dou, S. Study on Friction Characteristics of AA7075 Aluminum Alloy under Pulse Current-Assisted Hot Stamping. *Metals* 2023, *13*, 972; https://doi.org/10.3390/met13050972.
7. Teng, H.; Xia, Y.; Pan, C.; Li, Y. Modified Voce-Type Constitutive Model on Solid Solution State 7050 Aluminum Alloy during Warm Compression Process. *Metals* 2023, *13*, 989; https://doi.org/10.3390/met13050989.
8. Kang, J.; Cui, Y.; Zhong, D.; Qiu, G.; Lv, X. A New Method for Preparing Titanium Aluminium Alloy Powder. *Metals* 2023, *13*, 1436; https://doi.org/10.3390/met13081436.
9. Zhang, L.; Li, J.; Zhang, J.; Liu, Y.; Lin, L. First-Principle Investigation into Mechanical Properties of Al6Mg1Zr1 under Uniaxial Tension Strain on the Basis of Density Functional Theory. *Metals* 2023, *13*, 1569; https://doi.org/10.3390/met13091569.
10. Cao, P.; Xie, G.; Li, C.; Zhu, D.; Feng, D.; Xiao, B.; Zhao, C. Investigation of the Quenching Sensitivity of the Mechanical and Corrosion Properties of 7475 Aluminum Alloy. *Metals* 2023, *13*, 1656; https://doi.org/10.3390/met13101656.

References

1. Feng, D.; Li, X.D.; Zhang, X.M.; Liu, S.D.; Wang, J.T.; Liu, Y. The novel heat treatments of aluminium alloy characterized by multistage and non-isothermal routes: A review. *J. Cent. South Univ.* **2023**, *30*, 2833–2866. [CrossRef]
2. Abolfazl, A.; Ali, K.T.; Kourosh, K.T. Recent advances in ageing of 7xxx series aluminum alloys: A physical metallurgy perspective. *J. Alloys Compd.* **2019**, *781*, 945–983.
3. Zhao, X.M.; Meng, J.R.; Zhang, C.; Wei, W.; Wu, F.F.; Zhang, G.G. A novel method for improving the microstructure and the properties of Al-Si-Cu alloys prepared using rapid solidification/powder metallurgy. *Mater. Today Commun.* **2023**, *35*, 105802. [CrossRef]
4. Beder, M.; Alemeag, Y. Influence of Mg addition and T6 heat treatment on microstructure, mechanical and tribological properties of Al−12Si−3Cu based alloy. *Trans. Nonferrous Met. Soc. China* **2021**, *31*, 2208–2219. [CrossRef]
5. Beroual, S.; Boumerzoug, Z.; Paillard, P.; Yann, B.P. Effects of heat treatment and addition of small amounts of Cu and Mg on the microstructure and mechanical properties of Al-Si-Cu and Al-Si-Mg cast alloys. *J. Alloys Compd.* **2019**, *784*, 1026–1035. [CrossRef]
6. Ye, J.; Pan, Q.L.; Liu, B.; Hu, Q.; Qu, L.F.; Wang, W.Y.; Wang, X.D. Influences of small addition of Sc and Zr on grain structure and quenching sensitivity of Al-Zn-Mg-Cu alloys. *Mater. Today Commun.* **2023**, *35*, 1–9. [CrossRef]
7. Fang, Z.Y.; Krishanu, R.; Lim, J.B.P. Structural Design for Roll-Formed Aluminium Alloy Perforated Channels Subjected to Interior-Two-Flange Web Crippling: Experimental Tests, Numerical Simulation, and Neural Network. *Int. J. Steel Struct.* **2023**, *23*, 692–708. [CrossRef]
8. Kverneland, A.; Hansena, V.; Thorkildsen, G.; Larsen, H.B.; Pattison, P.; Li, X.Z.; Gjønnes, J. Transformations and structures in the Al−Zn−Mg alloy system: A diffraction study using synchrotron radiation and electron precession. *Mater. Sci. Eng. A* **2011**, *528*, 880–887. [CrossRef]

Disclaimer/Publisher's Note: The statements, opinions and data contained in all publications are solely those of the individual author(s) and contributor(s) and not of MDPI and/or the editor(s). MDPI and/or the editor(s) disclaim responsibility for any injury to people or property resulting from any ideas, methods, instructions or products referred to in the content.

MDPI
St. Alban-Anlage 66
4052 Basel
Switzerland
www.mdpi.com

Metals Editorial Office
E-mail: metals@mdpi.com
www.mdpi.com/journal/metals

Disclaimer/Publisher's Note: The statements, opinions and data contained in all publications are solely those of the individual author(s) and contributor(s) and not of MDPI and/or the editor(s). MDPI and/or the editor(s) disclaim responsibility for any injury to people or property resulting from any ideas, methods, instructions or products referred to in the content.

www.ingramcontent.com/pod-product-compliance
Lightning Source LLC
LaVergne TN
LVHW070640100526
838202LV00013B/844